Legionellosis

Volume II

Editor
Sheila Moriber Katz, M.D.
Professor
Department of Pathology and Laboratory Medicine
Hahnemann University School of Medicine
Philadelphia, Pennsylvania

CRC Press
Taylor & Francis Group
Boca Raton London New York

CRC Press is an imprint of the
Taylor & Francis Group, an **informa** business

First published 1985 by CRC Press
Taylor & Francis Group
6000 Broken Sound Parkway NW, Suite 300
Boca Raton, FL 33487-2742

Reissued 2018 by CRC Press

Library of Congress Cataloging in Publication Data
Main entry under title:

Legionellosis.

Includes bibliographies and index.
1. Legionnaires' disease. 2. Legionella pneumophila
I. Katz, Sheila Moriber, 1943- . [DNLM:
1. Legionnaires' Disease. WC 200 L5135]
RC152.7.L43 1985 616.9'2 85-416
ISBN 0-8493-5233-9 (v. 1)
ISBN 0-8493-5234-7 (v. 2)

A Library of Congress record exists under LC control number: 85000416

ISBN 13: 978-1-315-89488-1 (hbk)
ISBN 13: 978-1-351-07398-1 (ebk)

Visit the Taylor & Francis Web site at http://www.taylorandfrancis.com and the
CRC Press Web site at http://www.crcpress.com

PREFACE

Let's keep our watch. No matter what, we are never done discovering new agents or new ailments. Enter Legionella, bicentennial ruffian: it was Philadelphia, 1976, and bacteriology was presumed stagnant. Now the saga of a new microbe and a new disease unwrapped like a bandage, erupted like a full-scale battle, attesting that to be unseen is not to be unreal. Unfocused by lenses of modern science, Legionella had remained camouflaged before 1976 primarily by virtue of two iconoclastic traits: (1) it rejected conventional stains that visualize microbes and (2) its culture required a hitherto unconcocted medium. So, prior to its isolation in January, 1977[1] the elusive microbe hid in infected tissues and it died on conventional bacteriologic media. (No matter that someone had spied it in the lung of an ill Legionnaire in August, 1976).[2]

In the summer-fall of 1976, all tests confirmed the absence of a microbe. The medical community surmized that an infectious agent could not be there and nickle, paraquat, or freon were presumed the likely culprits of the Philadelphia outbreak. The scientists who finally cracked the mystery, particularly those diligent and dedicated workers at the Federal Centers for Disease Control, are, indeed, modern heros and heroines. They kept the circuitry open and they kept their watch. I am one eyewitness to their stamina, perserverance, and logic. The book in your grasp began with their collective efforts.

Legionellosis is a text in two volumes that presents the modern viewpoint of the agent and the disease. It also chronicles the history of the discovery of *Legionella pneumophila*. You will, for instance, meet Dr. Joseph McDade, the scientist who isolated Legionella, and you will read an account of the epidemic of Philadelphia, 1976. Volume I discusses current aspects of the microbe including taxonomy, morphology, biochemistry, and physiology. It also discusses the illness including clinical features, pathology, and therapy. Volume II details the laboratory diagnosis, epidemiology, and pathobiology. The contributors are amongst the most eminent scientists in their respective fields. To that end, thanks go to Dr. Albert Balows, Assistant Director, Laboratory Service Center for Infection Disease Federal Centers for Disease Control, for his efforts. Thanks are also offered to Dr. Jay M. Hammel, Associate Professor of Microbiology and Immunology, Hahnemann University, for his editorial assistance.

Please remember that even as you read this gospel, new data ebbs and flows. It is all so fluid, the unraveling of natural law.

Sheila Moriber Katz

REFERENCES

1. **McDade, J. E., Shepard, C. C., Fraser, D. W., et al.,** Legionnaires' disease: isolation of a bacterium and demonstration of its role in other respiratory disease, *N. Engl. J. Med.*, 297, 1197, 1977.
2. **Katz, S. M., Brodsky, I., and Kahn, S. B.,** Legionnaires' disease: ultrastructural appearance of the agent in a lung biopsy specimen, *Arch. Pathol. Lab. Med.*, 103, 261, 1979.

Legionella

Fin
below billows.

Strange vibrations,
Low, disharmonious.
But it is nothing, no one.
Only a hungry gull

Or a hull hauling ripples
To shore.
Still and sterile
The water is febrile,

Airless.
It is a lung fit for drowning.
Waves contuse blue to mauve to red,
And lips blench colorless.

Something below grinds alveoli.
Alien smells prickle the snout or the palate.
Hell of cries,
Hill of eyes,

They plunk like coral or legs
Into mud. Mauled:
The skulls and organs
Fix or float, cluttered in formalin.

Still no clarity, no cleaver.
?Are they slitted gills or lateral cuts,
These neck marks
Predict them.

All teeth, it is a man-eater,
Jaws of tiny grindstones or enzymes,
Each tooth is white, triangular.
No repellent

As the profile clarifies
To placoid scales or jellied membranes.
Part microbe, part Pisces,
A killer pure as atoms or love.

Be it fish or cell
The portrait develops
until
Fin and all,

It is netted —
Legionella.

Sheila Moriber Katz

THE EDITOR

Sheila Moriber Katz, M.D. is professor of Pathology and Laboratory Medicine at Hahnemann University School of Medicine and Director of the Division of Renal Pathology and Electron Microscopy. Dr. Katz received her B.A. from Cornell University College of Arts and Sciences and her M.D. degree from Duke University School of Medicine. Her postdoctoral training in pathology took place at Yale, Harvard, Jefferson, and the University of Pennsylvania. From 1974 to the present she has been on the faculty at Hahnemann, rising from the rank of assistant professor to full professor in 1981. Dr. Katz is a member of more than 20 professional societies and the author of more than 100 scientific publications. Her research interests include Legionellosis, renal diseases, and diagnostic electron microscopy.

CONTRIBUTORS

Charles E. Benson, Ph.D.
Associate Professor of Microbiology
Head, Large Animal Diagnostic Microbiology
School of Veterinary Medicine
New Bolton Center
University of Pennsylvania
Kennett Square, Pennsylvania

Baruch S. Blumberg, M.D.
Institute for Cancer Research
Fox Chase Cancer Center
Philadelphia, Pennsylvania

Amedo Bondi, Ph.D.
Professor
Department of Microbiology and Immunology
Hahnemann University School of Medicine
Philadelphia, Pennsylvania

Claire V. Broome, M.D.
Chief, Special Pathogens Branch
Bacterial Diseases Division
Center for Infectious Diseases
Centers for Disease Control
Atlanta, Georgia

Donald C. Drake
Staff Writer
Philadelphia Inquire
Philadelphia, Pennsylvania

Toby K. Eisenstein
Associate Professor
Department of Microbiology and Immunology
Temple University School of Medicine
Philadelphia, Pennsylvania

James C. Feeley, Ph.D.
Chief
Field Investigations Section
Respiratory and Special Pathogens Branch
Centers for Disease Control
Atlanta, Georgia

Carl B. Fliermans, Ph.D.
Staff Microbial Ecologist
Savannah River Laboratory
E.I. duPont deNemours and Company
Aiken, South Carolina

Herman Friedman
Chairman and Professor
Department of Medical Microbiology and Immunology
University of South Florida
School of Medicine
Tampa, Florida

Richard W. Gilpin, Ph.D.
Associate Professor
Department of Microbiology and Immunology
Medical College of Pennsylvania
Philadelphia, Pennsylvania

Marcus A. Horwitz, M.D.
Professor
Department of Medicine and Microbiology and Immunology
Chief
Division of Infectious Diseases
UCLA School of Medicine
Center for the Health Sciences
Los Angeles, California

Ellen J. Mangione, M.D.
Epidemic Intelligence Service
Bacterial Diseases Division
Center for Infectious Diseases
Centers for Disease Control
Atlanta, Georgia

Roger M. McKinney, Ph.D.
Supervisory Research Chemist
Biotechnology Branch
Centers for Disease Control
Atlanta, Georgia

Sheila Moriber Katz, M.D.
Professor
Department of Pathology and Laboratory Medicine
Hahnemann University School of Medicine
Philadelphia, Pennsylvania

Burton J. Landau, Ph.D.
Associate Professor of Microbiology and Immunology
Department of Microbiology and Immunology
Hahnemann University School of Medicine
Philadelphia, Pennsylvania

Joseph P. Matus
Department of Pathology
Hahnemann Medical College and Hospital
Philadelphia, Pennsylvania

Hazel W. Wilkinson, Ph.D.
Chief
Immunology Section
Centers for Disease Control
Atlanta, Georgia

ACKNOWLEDGMENT

To Sara, Jonathan, and Julian.

TABLE OF CONTENTS

Volume I

Part I: History of Legionellosis

Volume II

Part I: Laboratory Diagnosis of Legionellosis

Part II: Epidemiology of Legionellosis

Part III: Pathobiology of Legionellosis

Part IV: The New Microbe: Commentary

Part I: Laboratory Diagnosis of Legionellosis

Chapter 1

DIAGNOSIS: CULTURE

James C. Feeley

TABLE OF CONTENTS

I. INTRODUCTION

The diagnosis of acute cases of Legionnaires' disease has been primarily made clinically. Laboratory confirmation followed days or weeks later. However, this is changing with the development of rapid antigen detection tests[9] and the improvement of the culture media by addition of ACES buffer,[11] α-ketoglutarate,[4] and effective antimicrobials as selective agents[1,2,4,13] to the charcoal-yeast extract (CYE) base medium.[7] In the U.S., the sensitivity of culture currently ranges between 50 to 80% in laboratories that routinely culture specimens for *Legionella*. Microbiologists in these laboratories can detect *Legionella* as early as 2 days of incubation following inoculation of the media by picking microcolonies using a dissecting microscope, and staining them with immunofluorescent antibody reagents.

II. DEVELOPMENT OF MEDIA

Since an understanding of the growth requirements of *Legionella* species has proven helpful to many microbiologists in isolating *Legionella* species, a brief history of the development of the media is provided. The Legionnaires' disease bacterium (LDB), now called *L. pneumophila,* was first isolated by using rickettsial techniques.[10] Guinea pigs were inoculated intraperitoneally with lung tissue from fatal cases of Legionnaire's disease. When a Guinea pig became ill, its spleen was removed, ground, and suspended in sterile phosphate buffered saline and inoculated into yolk sacs of embryonated hens eggs. The small Gram-negative rods in the yolk sacs that were identified as the etiologic agent of Legionnaires' disease by indirect immunofluorescent staining grew only on Mueller-Hinton supplemented with 1% hemoglobin and 1% IsoVitalex® (MH-IH) agar but did not grow on 16 other types of media.[5] Feeley et al. examined components of the MH-IH medium[5] to determine what special requirements that LDB needed for growth.

They found that MH-IH agars made without L-cysteine HCL and iron salts were unable to support the growth of LDB, that pH 6.9 was optimal for "growth initiation" for LDB on primary isolation, and that cultures maintained on bacteriological media were less affected by pH. These observations have been applied in the development of most of the media designed specifically to grow *L. pneumophila* and in the development of quality control testing procedures for *Legionella.*

Feeley-Gorman (F-G) agar was the first medium that was specifically formulated to grow *L. pneumophila.* It was substantially more sensitive than MH-IH agar, and it also enabled detection of the characteristic browning that most *Legionella* species produce on tyrosine containing media. Unfortunately, the Mueller-Hinton basal medium used in making F-G agar varied widely from lot to lot in its ability to support growth of *L. pneumophila.* For this reason and the fact that F-G agar does not support primary isolation of non-*pneumophila Legionella* species, it should not be used for primary isolation. CYE agar was developed to replace F-G agar.[7] It was found to be 100 times more sensitive, to support faster growth, and to support primary isolation of all *Legionella* species. However, one shortcoming that CYE agar was found to have was its inability to recover *Legionella* from specimens such as sputa that contain other flora. One possible explanation for this was that the other flora produce substances such as bacteriocins that were inhibitory to *Legionella.* Another explanation was that the other flora rapidly changed the pH of the agar from 6.9 before *Legionella* could initiate growth. The addition of ACES buffer[11] was added to correct this fault by increasing the buffering capacity of CYE agar. The new medium that resulted was called "buffered charcoal yeast extract" (BCYE) agar and did support better growth of *Legionella* from all kinds of specimens. Another improvement of the basal medium resulted with the addition of α-ketoglutarate[4] to BCYE agar. This enabled faster initiation of growth of *Legionella* as well as giving higher recovery of *Legionella* from antibiotic containing media supplemented with it as contrasted to lower recovery of *Legionella* from antibiotic containing media not supplemented with it. BCYE media containing α-ketoglutarate is designated BCYEα agar.

III. RECOMMENDED MEDIUM

BCYEα agar and BCYEα diphasic blood culture medium are the best basal media currently available for culturing *Legionella.* BCYEα agar can be made selective by supplementation with various antimicrobials.[1,2,4,13] Some investigators have added dyes[12,13] to enable detection and differentiation of *Legionella* species. This has proved extremely helpful in examining water from hospitals contaminated with two or more species of *Legionella* such as *L. pneumophila* and *L. micdadei*. However, dye incorporation is not recommended for general usage because it complicates the medium and may inhibit some species of *Legionella*. Many investigators prefer using DFA staining for detecting the numerous species that may be in a specimen. Many commercial media companies are now selling BCYEα agar with and without antimicrobial and dye supplementation.

A. BCYEα Medium

1. Composition

Norit SG charcoal	2.0 g
Yeast extract	10.0 g
ACES buffer[a]	10.0 g
Potassium α-ketoglutarate	1.0 g
Ferric pyrophosphate soluble	0.25 g
L-Cysteine — HCL-H_2O	0.45 g
Agar	15.0 g
Distilled water	1.0 ℓ

[a] *N*-2-Acetamido-2-aminoethane-sulfonic acid.

2. Sources for Ingredients

All of the ingredients of BCYEα agar are now commercially available and most are easily obtained, including Norit SG charcoal and ferric pyrophosphate soluble. Sigma Chemical Co., St. Louis, Mo., sells both, which are listed, respectively, as products C5510 and P6526. Ferric pyrophosphate soluble can now also be purchased from City Chemical Corp., 132 West 22nd St., New York, N.Y. 10011, and is listed as product number F454. Although some agars have been found toxic to *Legionella;* agars from DIFCO and GIBCO and OXOID have been found to be suitable for use and should be added in amounts (grams per liter) suggested by the individual manufacturer.

3. Directions for Preparation

The yeast extract and some of the other components of BCYEα agar can be easily denatured during preparation of the medium. The following directions are provided to prevent this:

1. Add 940 mℓ of distilled water to 10 g ACES buffer and dissolve by warming in a 50°C water bath. Since the buffer does not come to its pKa of 6.9 and is very acid, it needs to be adjusted to pH 6.9 by adding approximately 40 to 43 mℓ of 1 *N* KOH to the buffer solution. Note well that NaOH must not be used to adjust the ACES buffer because it is inhibitory to *L. pneumophila* in the final medium.
2. Pour the buffer solution into a second flask that contains the dry powders of charcoal, yeast-extract, agar-agar, and α-ketoglutarate, and mix well.

3. Gently heat the mixture to dissolve the agar-agar and sterilize it by autoclaving at 15 lb for 15 min. The medium must not be subjected to prolonged heating because it is extremely sensitive to high temperatures.
4. Cool the medium immediately to 50°C by placing the flask in a 50°C water bath.
5. For medium intended to be *complete* (that containing L-cysteine), add 10 mℓ of a filter sterilized solution of L-cysteine (0.45 g/10 mℓ) and mix thoroughly. L-Cysteine should not be added to media that will be used to test L-cysteine requirement of cultures. This medium should be labeled "C-" for "Cysteine minus" BCYE agar.
6. Add separately 10 mℓ of a filter-sterilized solution of soluble ferric pyrophosphate (0.25 g/10 mℓ), and mix thoroughly. The iron and cysteine solutions should not be mixed prior to addition to medium because a toxic complex may result.
7. The final pH of the medium should be 6.90 ± 0.05 at 20°C. The pH of the final medium can be adjusted while the medium is being held at 50°C in the water bath. At this temperature the medium should be adjusted to 6.30. The reason is that the pKa of ACES buffer changes −0.02 pH units per +1°C greater than 20°C. On cooling to room temperature, the pH of the medium will change and remain at 6.90 even on incubation of cultures at 35°C.
8. For preparation of BCYEα + antibiotics agar, add membrane-filtered antibiotics and mix.
9. Dispense 20 mℓ or more per 15 × 100 mm petri dish. The medium must be swirled between the pouring of each plate to keep the charcoal particles suspended. After the media has solidified the plates should be stored in plastic bags in the refrigerator in the dark. They should be good for approximately 2 months provided that they passed quality control.

4. Precautions

The soluble ferric pyrophosphate must be kept dry and in the dark. If its green color becomes yellow or brown, it is not suitable for use.

5. Quality Control

Each batch of media must be tested for pH (6.90 ± 0.05) and ability to support the growth of *L. pneumophila*. Ideally, all media should be tested with a standardized tissue inoculum. However, since this material is not generally available, stock culture inoculum is recommended with the understanding that it is a minimal quality control test and can be done in the following manner:

1. Prepare a standardized inoculum by emulsifying some actively growing (in logarithmic stage) *L. pneumophila* in sterile distilled water to a turbidity of a McFarland #2 standard.
2. Seed the media with 0.05 mℓ of the standard inoculum. Streak the plates for isolated colonies, and incubate at 35°C in humidified air.
3. Examine the media daily. On agar plates, growth should be present in the heavily inoculated area after 1 day. Isolated colonies should be microscopically visible in 2 to 3 days. If cells in standard inoculum have been refrigerated or frozen, growth will be slower. Media that does not support growth of isolated colonies should not be used for primary isolation.

6. Antibiotic Supplementation

Several antibiotic supplementations have been developed.[1,2,4,13] However, no single one is effective for all clinical specimens. Since different supplementations inhibit different flora, use of more than one supplementation is helpful. Two formulations that the author has used with success are those reported by Edelstein:[2,4]

EDELSTEIN 1981 FORMULATION

Anisomycin	80 μg/mℓ[a]
Polymyxin B	80 units/mℓ
Cefamandole	4 μg/mℓ

[a] Components are listed as final concentrations in medium.

EDELSTEIN MODIFIED 1979 FORMULATION

Vancomycin	0.5 μg/mℓ[a]
Polymyxin B	40 units/mℓ
Anisomycin	80 μg/mℓ

[a] Components are listed as final concentrations of components in the medium.

B. BCYEα Diphasic Blood Culture Medium

1. Composition

SOLID PHASE

Activated charcoal (Norit A or Norit SG)	2.0 g
Agar (DIFCO)	17.0 g
Distilled water	500 mℓ

LIQUID PHASE

Yeast extract	20.0 g
ACES buffer	10.0 g
Potassium α ketoglutarate	1.00 g
L-Cysteine HCl-H_2O	0.40 g
Fe $(NO_3)_3$-$9H_2O$	0.10 g
Sodium polyanetholsulfonate	0.25 g
Distilled water	500 mℓ

2. Preparation

The solid-phase is prepared first. Ingredients are combined, boiled, and dispensed at 20-mℓ aliquots into each of a series of 125-mℓ Wheaton serum bottles which are loosely stoppered with both a rubber stopper and a metal cap. The bottles are then autoclaved at 121°C for 20 min. When the bottles cool to 50°C, the warm charcoal and agar are thoroughly remixed and the bottles were placed at an angle so that an agar slant with a vertical height of 6.0 cm is formed. This should result in a portion of the agar protruding above the 25 mℓ of liquid-broth phase after the blood is added.

The liquid phase is prepared by first autoclaving the yeast extract and water at 121°C for 15 min. When it cooled to <50°C, fresh sterile solutions of L-cysteine HCl and ferric nitrate are added in that order. After the pH is adjusted to 6.9 by adding ~6.0 mℓ of 1 *N* KOH, sterile sodium polyanetholsulfonate (liquid) obtained as Grobax® from Hoffman-La-Roche, Inc. is added to the broth to give a 0.025 final concentration (vol/vol) in the complete CYE-diphasic medium. Aliquots of this (20 mℓ) are dispensed into the Wheaton bottles containing charcoal agar slants. The bottles were sealed by crimping the metal caps over the rubber stoppers. Each bottle should be checked for sterility by preincubation at 35°C for at least 2 days.

IV SPECIMENS

A. Safety Considerations

Since Legionnaires' disease is primarily a pulmonary infection, the prevention and containment of aerosols is essential. Routine precautions can be taken to assure this in the collection and processing of human specimens. However, individuals collecting and processing infected animal specimens should wear protective clothing, (gown, mask, and gloves) and practice good bacteriological technique.

B. Types

In the past, *Legionella* were only able to be recovered mainly from respiratory specimens that were free or had few other bacteria because they prevented *Legionella* from growing. This is no longer true with the current selective media. Culturing of sputa is recommended and can have a sensitivity as great as 80% or greater in recovery of *Legionella* from documented cases of legionellosis. Although there have been few reports of isolation from blood, this is due to lack of culturing. An effective diphasic medium[6] is available and culturing of blood is recommended.

C. Collection and Transport

Specimens for culture should be collected and transported in sterile, leakproof containers directly to the laboratory and should be cultured within 2 hr of collection. When this is not possible, specimens can be refrigerated but no longer than 4 days. If culturing will be delayed longer than this, specimens should be quick-frozen and stored at −70°C. In addition, the original specimens should be refrigerated for further processing if the plates are overgrown with contaminating flora. This can be seen usually by the next day. Specimens then can be either acid treated or heated at 60°C for 1 to 2 min before being reinoculated to plate media.

V. CULTURE AND PRIMARY ISOLATION

Blood specimens should be inoculated (5 mℓ) directly into diphasic medium, incubated at 35°C, and examined daily for turbidity with subsequent growth on the slant. Subcultures should be made to BCYEα media and examined as described in the following section. Other liquid specimens such as pleural fluid usually can be inoculated directly onto media. However, specimens with greater consistency such as lung tissue should be emulsified in sterile water with a mortar. In order to neutralize any tissue toxicity, an aliquot of each specimen should be diluted at least 1:100 in sterile water and inoculated with the undiluted specimen to one plate each of BCYE agar with and without antibiotics. Streak plates for isolated colonies with a sterile bacteriological loop and incubate at 35°C in a moist chamber which contains air or air plus no more than 2.5% CO_2. Since most species of *Legionella* except *L. gormanii* do not require CO_2 for primary growth, CO_2 supplementation is not generally recommended unless *L. gormanii* is suspected.

VI. EXAMINATION AND IDENTIFICATION OF CULTURES

Colonies of *Legionella* species can appear on BCYEα agar as early as 2 days, but they generally appear after 3 to 5 days, and on occasion appear only after 2 weeks of incubation. Therefore, plates should be incubated for at least 2 weeks before being discarded. Young colonies of *Legionella* are very distinctive when obliquely illuminated at a 45° angle with an incandescent or fluorescent lamp and observed with a dissecting microscope. They appear as pinpoint blue dots with a cut-glass texture. Since this characteristic is usually lost within a day or so, it is recommended that plates be examined daily in order to detect the colonies when they are very young. Older colonies are larger, smooth, and appear gray-tan. On subculture both large and small colonies result giving the appearance of a mixed culture which is not the case. Long-wave UV light (366 nm) can be used to advantage in specimens having both blue-white fluorescing and nonfluorescing species. Three fluorescent and three nonfluorescent colonies (if present) should be subcultured to BCYEα agar prepared with and without L-cysteine HCl. Sufficient growth should occur to be visible after overnight incubation at 35°C. Culture that grow only on medium with L-cysteine HCl should be considered as possible *Legionella*. Although two new species of *Legionella* have been reported that appear not to require L-cysteine on subculture, they are environmental isolates that have not been isolated to date clinically. Cultures suspected of being *Legionella* should be examined by DFA staining and also examined for biochemical activity. See appropriate chapters in this book for specific details.

REFERENCES

1. **Bopp, C. A., Summer, J. W., Morris, G. K., and Wells, J. G.,** Isolation of *Legionella* spp. from environmental water samples by low pH treatment and use of a selective medium, *J. Clin. Microbiol.*, 13, 714, 1980.
2. **Edelstein, P. H. and Finegold, S. M.,** Use of a semiselective medium to culture *Legionella pneumophila* from contaminated lung specimens, *J. Clin. Microbiol.*, 10, 141, 1979.
3. **Edelstein, P. H., Meyer, R. D., and Finegold, S. M.,** Isolation of *Legionella pneumophila* from blood, *Lancet*, 1, 750, 1979.
4. **Edelstein, P. H.,** Improved semiselective medium for isolation of *Legionella pneumophila* from contaminated clinical and environmental specimens, *J. Clin. Microbiol.*, 14, 298, 1981.
5. **Feeley, J. C., Gorman, G. W., Weaver, R. E. et al.,** Primary isolation media for the Legionnaires' disease bacterium, *J. Clin. Microbiol.*, 8, 320, 1978.
6. **Feeley, J. C., Gorman, G. W., and Gibson, R. J.,** Primary isolation media and methods, in *"Legionnaires:" the Disease, the Bacterium and Methodology*; Jones, G. L. and Hebert, G. A., Eds., Centers for Disease Control, Atlanta, Ga., 1979, 77.
7. **Feeley, J. C., Gibson, R. J., Gorman, G. W. et al.,** CYE agar: a primary isolation medium for the Legionnaires' disease bacterium, *J. Clin. Microbiol.*, 10, 436, 1979.
8. **Greaves, P. W.,** New methods for the isolation of *Legionella pneumophila*, *J. Clin. Pathol.*, 33, 581, 1980.
9. **Kohler, R. B., Zimmerman, S. E., Wilson, E. et al.,** Rapid radioimmunoassay diagnosis of Legionnaires' disease: detection and partial characterization of urinary antigen, *Ann. Intern. Med.*, 94, 601, 1981.
10. **McDade, J. E., Shepard, C. C., Fraser, D. W. et al.,** Legionnaires' disease. Isolation of a bacterium and demonstration of its role in other respiratory diseases, *N. Engl. J. Med.*, 297, 1197, 1977.
11. **Pasculle, S. W., Feeley, J. C., Gibson, R. J. et al.,** Pittsburgh pneumonia agent: direct isolation from human lung tissue, *J. Infect. Dis.*, 141, 727, 1980.
12. **Vickers, R. M., Brown, A., and Garrity, G. M.,** Dye-containing medium for differentiation of members of the family *Legionellaceae*, *J. Clin. Microbiol.*, 13, 380, 1981.
13. **Wadowsky, R. M. and Yee, R. B.,** Glycine-containing selective medium for isolation of Legionellaceae from environmental specimens, *Appl. Environ.*, 42, 768, 1981.

Chapter 2

DIAGNOSIS: SEROLOGY

Hazel W. Wilkinson

TABLE OF CONTENTS

I. DETECTION OF ANTIBODY

A. Immune Response to Infection

Available data suggest that patients with legionellosis produce antibodies of the immunoglobulin (Ig) G, M, and A classes, singly or in various combinations.[1,2] No single Ig class appears to be a more specific indicator of *Legionella* infection or of the interval of time since onset of infection, with the possible exception of IgA, which declines more rapidly but which is detectable less frequently than IgG or IgM. In most patients with Legionnaires' disease (LD) antibodies are detectable 3 to 6 weeks after onset of symptoms,[3] and both IgG and IgM can persist for months after the initial episode, depending upon the magnitude of the peak titer.[4,36] A followup study of patients from the 1976 Legionnaires' disease outbreak[5] suggested that indirect immunofluorescence assay (IFA) titers declined at a rate of approximately one twofold dilution factor per year.[37] Data in Table 1 show that convalescent-phase IFA titers against an antigen of the same species and serogroup as the infecting strain can be many-fold higher than the corresponding acute-phase titers, even for patients given immunosuppressive drugs or who are immunocompromised in other ways. They also suggest that titers as high as 256 may not be protective.

Data in Table 1 and past studies[1,2,4,6] also show that antibodies with multiple specificity are produced in the majority of patients with legionellosis. In sera studied so far from patients with positive culture or direct immunofluorescence assay (DFA) test results, titers against an antigen of the same serogroup as the infecting strain were usually at least as great as or, more often, greater than those against heterologous *Legionella* antigens. This can probably be attributed to the additive effect of having both serogroup-specific and common antigen-reactive antibodies in these patients' sera. The distribution of positive IFA titers against 29 *Legionella* antigens among 444 patients with suspected Legionnaires' disease suggests that human antibodies can react with numerous antigenic determinants on the surface of *Legionella* cells.[7] Titers of sera from an individual patient could be detected at the diagnostic level (for *L. pneumophila*, seroconversion to $\geq$128) against from 1 to 13 different serogroups or species antigens. At least a dozen patterns of interspecies cross-reactivity were noted and, within *L. pneumophila*, interserogroup cross-reactivity was observed among serogroups 1 and 4, 4 and 5, and 1 and 6 antigens. For these reasons, antibody tests cannot be used to identify the serogroup or species of the infecting strain, although they can be used to diagnose *L. pneumophila* serogroup 1 infections with reasonable sensitivity and specificity. Whether this is true for infections caused by other serogroups and species cannot be determined until sufficient specimens from well-documented cases are available for testing.

That patients with legionellosis also produce antibodies reactive with antigenic determinants common to *Legionella* and several other Gram-negative bacterial species was shown in studies using *Legionella*, *Escherichia*, *Haemophilus*, *Neisseria*, *Pseudomonas*, *Klebsiella*, and *Salmonella* strains as IFA antigens.[2] These antibodies could be blocked from reacting with the Gram-negative cross-reactive antigens with a hot aqueous extract of *E. coli* strain 013:K92:H4. Failure of purified K92 polysaccharide antigen (generously donated by Dr. John Robbins) to similarly inhibit antibody binding,[38] and the fact that the immunosorbent and IFA antigens were prepared from cells subjected to a boiling water bath suggest that the cross-reactions were not due to capsular or flagellar antigens. Instead, there may be common polysaccharide determinants in the lipopolysaccharide (LPS) of these organisms. They are so far unidentified, however, as only the LPS from *E. coli* 055 inhibited the cross-reactive antibodies in a sample of sera tested against *L. pneumophila* serogroup 1 and *E. coli* 013 antigens, but only at a concentration of 5 mg/mℓ. Concentrations of 1 mg/mℓ and 100 μg/mℓ of 055 LPS, and 5 mg/mℓ of LPS fom *E. coli* 0111, 0127, and 0128 failed to show similar inhibition to that obtained with the more concentrated suspension and with the *E. coli* immunosorbent. Surprisingly, immunosorbent did not substantially increase the

Table 1
ANTIBODY RESPONSE IN PATIENTS WITH LEGIONNAIRES' DISEASE

Case no.	Diagnosis confirmed by	*Legionella* identity species serogroup	Underlying condition	Antibody Response to Homologous Antigen: Days post-onset	IFA Titer	Ig Class (≥64)	Antibody to common antigen
1	Culture (lung)	*pneumophila* 1	Renal insufficiency; sarcoidosis	0 20	<64 4096	GMA	Yes
2	Culture (lung)	*pneumophila* 1		3 21	256 8192	GMA	Yes
3	DFA[a] (lung)	*pneumophila* 1	Renal transplant	−11 21	<64 ≥2048	Not determined	Yes
4	DFA (lung aspirate)	*pneumophila* 1	Heart transplant	2 37	<64 2048	GMA	ND[b]
5	DFA (sputum)	*pneumophila* 1		10 54	512 ≥2048	GMA	Yes
6	DFA (lung)	*pneumophila* 1	Systemic lupus erythematosus	4 12	64 512	GM	Yes
7	DFA (lung)	*pneumophila* 1	Renal transplant	−114,34 145	256 8192	GM	Yes
8	Culture (bronchial washing)	*pneumophila* 4	Hodgkin's disease	5 27	64 8192	GMA	Yes
9	Culture (lung)	*pneumophila* 6	Renal transplant	−4 29	128 32,768	GMA	Yes
10	Culture (lung)	*pneumophila* 6	Renal transplant	2 37	<64 32,768	GMA	Yes
11	Culture (blood)	*pneumophila* 6	Lymphoma	3 58	<64 256	GM	Yes
12	Culture (lung)	*bozemanii* 1		17[c]	<64 512	M	ND
13	DFA (lung)	*bozemanii* 1	Lymphoma; hepatitis; monoclonal gammopathy	5 25	<64 <64	ND	ND
14	DFA (sputum)	*longbeachae* 1		13 40	256 ≥2048	GA	Yes

[a] DFA, direct immunofluorescence assay. [b] ND, Not detected. [c] Interval between serum collection dates. Onset date unknown.

specificity when sera from patients in outbreaks of pneumonia were used to validate the IFA test.[1] Therefore, it is not used routinely in serodiagnostic testing.

B. Serologic Tests

1. Reference Test

The IFA is the method of choice for demonstrating seroconversions to *Legionella* antigens, and thus for providing retrospective serologic evidence of infection, as it is the only serologic test that has been standardized for this purpose.[8] The lack of sufficient numbers of sera from culture, or even DFA-confirmed legionellosis cases, has hampered the evaluation of alternative tests that might be advantageous because of their greater rapidity, simplicity, or lower cost. With more frequent isolations of *Legionella* and more widespread use of the DFA technique, this may become less of a problem in future studies. Table 2 provides a list of published methods, each of which will be discussed in general below. Specific methodology can be obtained from the original publications listed in the table.

An IFA developed with a strain isolated from a patient in the Philadelphia outbreak of 1976 was first used to demonstrate that patients with Legionnaires' disease show seroconversions during convalescence to *L. pneumophila* serogroup 1 antigens.[5] Of 111 patients for whom appropriately timed serum specimens were available, 90.9% demonstrated at least fourfold rises in IFA titer to $\geq$64 (n = 62), or single or sustained titers of $\geq$128 (n = 39). In contrast, serum specimens from 338 of 347 (97.4%) controls had IFA titers that were lower than those cutoff levels, including 11 Pennsylvania Legionnaires who did not attend the convention, 31 health department personnel, 178 of 182 patients whose sera were submitted to CDC for viral, *Mycoplasma*, Q fever, or psittacosis antibody testing but gave negative results in those tests, 108 of 113 patients whose sera gave positive titers in one of the latter tests, and 10 sera from patients with rheumatoid factor or antinuclear antibody titers. Cutoff levels for a positive test result were determined for each subsequent outbreak investigation with sera from epidemic-associated patients and appropriate controls, as the seroprevalence of antibody in healthy individuals varies considerably among different population groups in the U.S. (see below). This contributed to the decision to use cutoff levels that were one twofold dilution factor higher than those used for the Philadelphia investigation in testing sporadic case sera from patients through the U.S. Cutoff levels must be redefined if the incidence of disease changes or if any component of the test is modified, including the antigen (strain, growth conditions, diluent, killing, and fixation methods), the serum (human or animal, heat inactivation, diluent), the conjugate (class specificity, titer), the microscope (type of light, filters, magnification), and definition of fluorescence end point. The antigen has been the most frequently modified component of the test.

The original antigen was a 5% suspension of *L. pneumophila* serogroup-1-infected yolk sacs of embryonated hens' eggs.[5] As knowledge increased about the organism and as media were developed that could support its growth, the antigen was prepared from enriched Mueller-Hinton agar-grown cells in 1.0% normal yolk sac suspensions. Diethyl ether was used to kill the organisms for reference antigens, which by then were being distributed by CDC world-wide. When the Togus 1 strain of *L. pneumophila* serogroup 2 was prepared as an IFA antigen by this method, it was difficult to determine serum end points because the antigenic integrity of the cell surface had apparently been disrupted by ether extraction.[6] Killing could be accomplished instead by placing the cell suspension in a boiling water bath for 15 min. Heat-killed *L. pneumophila* serogroups 1 and 2 cells exhibited a more homogeneous pattern of staining than the ether-killed preparations and gave comparable IFA titers to the ones obtained with the preceding reference antigen (for serogroup 1) and with antigens prepared from live cells.

As media were improved for growing *Legionella* cells, antigen suspensions were prepared accordingly; from enriched Mueller-Hinton agar-, to FG agar-, and lastly, to CYE agar-grown cells. Each compared favorably with the preceding reference antigen, except the F-

Table 2
ANTIBODY DETECTION TESTS FOR LEGIONELLOSIS

Test method	Reactants	Ig Class detected	Advantages	Disadvantages	Ref.
Indirect immunofluorescence assay	LP 1-4[a] cells, serum, FITC-conjugate[c]	GMA	Reference test,[b] reproducible, relatively fast	Microscopic	15, 16
Immunofluorometric assay	LP 1 cells, serum, FITC-conjugate	GMA	Fluorometric readout, reproducible	Not standardized, lengthy protocol	12
Microagglutination	LP 1 cells, serum	M	Simple	Insensitive	10
Micro-enzyme-linked immuno-sorbent assay	LP 1 soluble antigen, serum, HP-conjugate,[d] substrate	GM (A?)	Amenable to spectrophotometric readout	Irreproducible, lengthy protocol	10
Indirect hemagglutination	LP 1-4 soluble antigen-erythrocytes, serum	GM	Simple	Not standardized	11
Solid-phase immunofluorescence assay	Sonicated LP 1 cells, serum, FITC-conjugate	GMA	Fluorometric readout, reproducible	Large volume of reactants required	13
Kinetic-enzyme-linked immuno-sorbent assay	LP 1 soluble antigen, serum, HP-conjugate, substrate	NT[e]	Semiautomated, reproducible, uses purified antigens	Equipment not available commercially	41

[a] LP, *Legionella pneumophila*. Numbers denote serogroup(s).
[b] Sensitivity = 78%, specificity = 99%, reproducibility = 98%.
[c] FITC, fluorescein isothiocyanate.
[d] HP, horseradish peroxidase.
[e] NT, not tested.

G agar-grown antigen which apparently had increased in sensitivity after repeated passages on F-G agar.[39] Reference serum titers were substantially higher against the laboratory-adapted strain than against the same strain that had been maintained in yolk sacs or grown on CYE agar. A test sensitivity of ≥78% and specificity of 99% were estimated when sera from patients with outbreak-associated cases of pneumonia were tested against the CYE agar-grown, heat-killed *L. pneumophila* serogroup 1 antigen (Philadelphia 1 strain). The reproducibility of the test was 98%. A polyvalent antigen containing *L. pneumophila* serogroups 1 to 4 gave comparable titers to those obtained with the constituent monovalent antigens.

In an early study at CDC, we compared yolk-sac-grown, formalin-killed antigen used in England with the CDC reference antigen.[40] If titer variations of ± 1 twofold dilution factor are considered in agreement, titers of 22 of 24 sera agreed when tested against the two antigens. Of the two discrepancies, 1 of 12 low-titered (≤64) sera had a titer of 64 against the yolk-sac-formalin antigen and ≤32 against the agar-heat (CDC) antigen, and 1 of 12 high-titered sera (≥128) had a titer of 128 against the yolk sac antigen and 512 against the CDC antigen. Discrepant titers can be the result of strain differences within the same serogroup, even when the same methods are used to prepare the antigens. Therefore, to eliminate as many variables as possible and to provide a data base against which to evaluate discrepant opinions regarding the relative merits of exposing *Legionella* cells to heat or to formalin, we compared these two methods of killing *L. pneumophila* serogroup 1 cells (Philadelphia 1 strain) that had grown on CYE agar.[9] This medium was used after we first determined that antigenic stability was maintained for at least five subpassages of yolk-sac-infected inoculum on CYE agar. Formalin at concentrations of 1 to 10% appeared to slightly disrupt the antigenic integrity of the cell surface, as previously observed with ether, and to cause IFA titers of human sera to be approximately one twofold dilution factor lower. However, if the cutoff level for a positive test result for the formalinized antigen was lowered by one tube from the levels used for the heated antigen, comparable results were obtained (96% agreement in testing paired sera from 60 patients). Both antigens bound antibodies with multiple specificity.

2. Alternative Tests

Tests other than the IFA that have been developed to measure antibodies to *Legionella* infection include the microagglutination test (MAT),[10] enzyme-linked immunosorbent assay (ELISA),[10] indirect hemagglutination test, (IHA),[11] immunofluorometric assay (IFMA),[12] solid-phase immunofluorescence assay (FIAX),[13] and kinetic-ELISA (K-ELISA)[41] (Table 2). There are advantages and disadvantages to each compared with the IFA. In addition to not requiring use of a microscope to determine test results, advantages include the technical simplicity of the MAT and IHA, the objectivity of titer determination with the IFMA and FIAX, the use of only one serum dilution with the ELISA, K-ELISA, IFMA, and FIAX, and the use of purified and (perhaps more specific) antigens in the ELISA and K-ELISA. The principal disadvantages of these alternative serologic tests are insufficient standardization (for reasons discussed above), lengthy test procedures for the ELISA and IFMA, inability to detect all immunoglobulin classes (MAT), and the requirement of large volumes of reagents relative to the IFA (FIAX). Tests that were standardized with normal control sera with low antibody levels from restricted populations (e.g., the IHA in Michigan and the IFMA in Georgia) require further validation with sera from diverse population groups.

C. Seroprevalence Studies

As shown in Table 3, the seroprevalence of antibodies against *L. pneumophila* serogroup 1 IFA antigen varies widely among different population groups. The proportion of 4035 non-ill individuals in the U.S. showing titers of ≥128 in 9 studies was 9.7%, with a range of from 1.3% in Atlanta to 19.9% in Ohio. In contrast, only 2.9% of 3831 normal control sera tested in Great Britain showed titers of ≥64. It is unlikely that differences in method

Table 3
LEGIONELLA PNEUMOPHILA SEROGROUP 1 TITERS IN NORMAL POPULATIONS

Location	No. of sera tested	% with IFA titer of: ≥64	≥128	≥256	Ref.
New York City	397		19.6	7.8	17
Scotland	1808	5.5			18
New York City	720		11.0		19
Iowa	517		13.2	4.5	20
Columbus, Ohio	322			0.6	21
Ohio	588		19.9		22
Washington, D.C.	199		1.5		23
Atlanta	304		1.3		23
Houston	406		1.7		23
Rochester	234		2.1		23
Nottingham, England[a]	2023	0.5	0.1		24
Los Angeles	116		4.3		25
Bristol	122		5.7		26
Burlington	110		15.5		27
Benidorm, Spain	21		19.0		28
12 U.S. states	184	42.4	16.3	3.8	7

[a] Titers against formalinized antigen.

Table 4
SEROPREVALENCE OF ANTIBODY AGAINST 29 *LEGIONELLA* ANTIGENS

Group	No.	% Showing IFA titer: SC[a] ≥ 128	≥128	≥256
Ill: 28 U.S. states	444	14.9		27.3
Non-ill: 12 U.S. states	184		40.8	12.0

[a] SC, seroconversion.

were solely responsible for the lower background titer levels in Britain, because titers obtained against formalinized antigens tend to be only one twofold dilution factor lower than titers obtained against heat-killed antigens (see above), and because the data from Scotland were obtained with the heat-killed antigen. The occurrence of elevated antibody titers is not restricted to the U.S., as shown by the fact that 19.0% of 21 sera from well individuals in Benidorm, Spain, had titers ≥128, whereas only 1.3% of 304 sera from well persons in Atlanta had IFA titers of ≥64. In the CDC study of sera from 12 states in the U.S. (Table 3), the proportion of sera with IFA titers ≥128 against *L. pneumophila* was 16.3%.

The proportion of sera from well individuals with positive IFA titers against antigens prepared from multiple *Legionella* species and serogroups was $2^1/_2$ (titers ≥128) to 3 times (titers ≥256) the proportion against only *L. pneumophila* antigens (Tables 3 and 4). Of 184 well individuals, 41% had titers of ≥128 against one or more of 29 antigens, including *L. pneumophila* serogroups 1 to 6, *L. bozemanii, L. dumoffii, L. gormanii, L. micdadei, L. longbeachae* serogroups 1 and 2, *L. jordanis,* and 8 additional *Legionella*-like organisms with undetermined taxonomic relationships; 12% of the sera had titers of ≥256: approximately half of these showed titers that were at least fourfold greater against one species antigen than against all of the others tested, but the remaining sera did not. Instead, multiple patterns of cross-reactivity occurred, suggesting that antibodies with multiple specificity reacted with antigenic determinants that are shared by most of the *Legionella* species. Titers in the well population could be the result of past infection or mere exposure to an expanding list of legionellae that are ubiquitous water inhabitants, or perhaps to cross-reactive antigens as yet unidentified.

Whether antibodies reactive with multiple *Legionella* species can also be detected in sera from patients with suspected Legionnaires' disease was determined by testing 444 paired or serial sera sent to CDC for *Legionella* IFA testing from 28 state health laboratories.[7] Of the serum sets, 15% showed $\geq$ fourfold increases in titer to $\geq$128 (seroconversion) from the acute to the convalescent-phase specimen against 1 or more of the 29 antigens, whereas only 6% showed seroconversion to the *L. pneumophila* antigen alone. As shown in tests of sera from normal individuals, approximately one half of the positive sera had titers that were at least fourfold greater against one species antigen than against all of the others tested. Within the *L. pneumophila* species, about half of the positive titers were fourfold or greater against only one serogroup antigen. Species- or serogroup-specific titers were the result of IgG and IgM, singly or combined, and remained elevated even when immunosorbent was used to block cross-reactive intergeneric antibodies. As found in previous studies with *L. pneumophila* serogroup 1 antigens, seroconversions were also detected for some of the non-*pneumophila Legionella* reactive sera against *M. pneumoniae* antigens in complement fixation (CF) tests (three of eight serum pairs).

Concomitant *Legionella* and *Mycoplasma* titers could not be interpreted with available data, as no cross-reactive antigens were demonstrated by staining ten *Mycoplasma* species with *L. pneumophila* conjugates,[42] and as concomitant antibody titers were not found in testing sera from patients with epidemic-associated Legionnaires' disease and *Mycoplasma* pneumonia.[1] Concomitant or secondary infections or polyclonal immune responses are possible explanations. However, *Mycoplasma* CF titers of an apparently nonspecific nature occurred in 41% of sera from patients with culture documented bacterial meningitis and 10% of sera from those with bacteremia of nonmycoplasmal etiology, an observation that called for caution in interpreting the *Mycoplasma* test.[14] Of the 80 serum sets obtained from patients with outbreak-associated cases of *Mycoplasma* pneumonia, psittacosis, Q fever, and tularemia and from patients with culture-confirmed bacteremia, 3 showed seroconversions against one or more *Legionella* antigens, for an estimated specificity of 96% for the multiple antigen *Legionella* IFA.[7]

Although the specificity of the IFA for testing sera from pneumonia patients against *L. pneumophila* antigens alone or for all 29 *Legionella* antigens can be estimated — 99 or 96%, respectively — the sensitivity can be estimated from currently available data only for *L. pneumophila* serogroup 1 (approximately 80%, 1), as sufficient numbers of epidemic-associated sera are available only for *L. pneumophila* serogroup 1 disease. There are insufficient numbers of sera from DFA- or culture-confirmed patients (Table 1) for sensitivity estimates or for determining the predictive value of a positive test result, which could range theoretically from 12 to 78% depending on the frequency of disease due to non-*pneumophila* legionellae.[7]

D. Recommendations

Currently, CDC tests sera referred from patients with suspected legionellosis against a heat-killed, polyvalent IFA antigen containing *L. pneumophila* serogroups 1 to 4. A seroconversion, demonstrated for appropriately timed paired sera (1 week pre- or post-onset and 3 to 6 weeks later) is considered evidence of a recent *Legionella* infection. Because of high seroprevalence rates in some population groups, we cannot interpret single or standing high titers as evidence of recent infection in sporadic cases. In outbreak investigations, the interpretation of a *Legionella* titer not showing a seroconversion relies on comparison with the appropriate controls. Although using multiple *Legionella* species antigens is often justifiable in well-controlled outbreak investigations, we also do not recommend including non-*pneumophila* antigens in routine serodiagnostic tests for legionellosis unless further data are obtained that would allow definition of the sensitivity of the IFA with these antigens and the predictive value of a positive test result.

II. DETECTION OF SOLUBLE ANTIGEN

As shown in Table 5, detection of *L. pneumophila* serogroup 1 soluble antigen in the urine of patients with Legionnaires' disease appears to be a highly specific (~100%) diagnostic aid and, depending upon the test method used, may also be quite sensitive (>90%). Obvious advantages to this approach include an earlier diagnosis than available by culture, DFA, or IFA techniques and specimen collection procedures that are noninvasive. Future studies should include modification of the antigen detection tests to include additional species and serogroups, and further evaluation of rapid, simple tests, such as the staphylococcal protein A-agglutination assays.

Table 5
ANTIGEN DETECTION TESTS FOR LEGIONELLOSIS

Test method	Reactants	Antigen detected	Urine specimens (cases)	Controls (no. positive/ no. tested)	Ref.
Enzyme-linked immunosorbent assay	Antiserum, urine, antiserum, AP-conjugate,[a] substrate	*L. pneumophila* serogroup 1	3/5	0/20	29
	Antiserum, urine, HP-conjugate,[b] substrate	*L. pneumophila* serogroup 1	1/1	0/4	30
	Antiserum, urine, AP-conjugate, substrate	*L. pneumophila* serogroup 1	39/47	0/178	31
Radioimmunoassay	Antiserum, urine, ^{125}I-conjugate	*L. pneumophila* serogroup 1	38/47	0/241	31, 32
Reversed passive hemagglutination	Antiserum-SRBC,[c] urine	*L. pneumophila* serogroup 1	1/1	NT[d]	33
Staphylococcal protein A agglutination assays	SAT: Soluble antigen, antiserum-*Staph.* cells	*L. pneumophila* serogroups 1—6, *L. bozemanii, L. dumoffii, L. gormanii, L. micadei*	NT	NT	34
	SIA: *Staph.* cells, antiserum, soluble antigen, FITC - conjugate[e]		NT	NT	34
	RPA: Urine, antiserum-*Staph.* cells		14/15	0/263	35

[a] AP, alkaline phosphatase.
[b] HP, horseradish peroxidase.
[c] SRBC, sheep erythrocytes.
[d] NT, not tested.
[e] FITC, fluorescein isothiocyanate.

REFERENCES

1. **Wilkinson, H. W., Cruce, D. D., and Broome, C. V.,** Validation of *Legionella pneumophila* indirect immunofluorescence assay with epidemic sera, *J. Clin. Microbiol.*, 13, 139, 1981.
2. **Wilkinson, H. W., Farshy, C. E., Fikes, B. J., Cruce, D. D., and Yealy, L. P.,** Measure of immunoglobulin G-, M-, and A-specific titers against *Legionella pneumophila* and inhibition of titers against non-specific, gram-negative bacterial antigens in the indirect immunofluorescence test for legionellosis, *J. Clin. Microbiol.*, 10, 685, 1979.
3. **Kirby, B. D., Snyder, K. M., Meyer, R. D., and Finegold, S. M.,** Legionnaires' disease: clinical features of 24 cases, *Ann. Intern. Med.*, 89, 297, 1978.
4. **McKinney, R. M., Wilkinson, H. W., Sommers, H. M., Fikes, B. J., Sasseville, K. R., Yungbluth, M. M., and Wolf, J. S.,** *Legionella pneumophila* serogroup six: isolation from cases of legionellosis, identification by immunofluorescence staining, and immunological response to infection, *J. Clin. Microbiol.*, 12, 395, 1980.
5. **McDade, J. E., Shepard, C. C., Fraser, D. W., Tsai, T. R., Redus, M. A., and Dowdle, W. R.,** Isolation of a bacterium and demonstration of its role in other respiratory disease, *N. Engl. J. Med.*, 297, 1197, 1977.
6. **Wilkinson, H. W., Fikes, B. J., and Cruce, D. D.,** Indirect immunofluorescence test for serodiagnosis of legionnaires disease: Evidence for serogroup diversity of Legionnaires disease bacterial antigens and for multiple specificity of human antibodies, *J. Clin. Microbiol.*, 9, 379, 1979.
7. **Wilkinson, H. W., Reingold, A. L., Brake, B. J., McGiboney, D. L., Gorman, G. W., and Broome, C. V.,** Reactivity of serum from patients with suspected legionellosis against 29 antigens of legionellaceae and *Legionella*-like organisms by indirect immunofluorescence assay, *J. Infect. Dis.*, 147, 23, 1983.
8. **Wilkinson, H. W.,** Serologic diagnosis of legionellosis, *Lab. Med.*, 13, 151, 1982.
9. **Wilkinson, H. W. and Brake, B. J.,** Formalin-killed versus heat-killed *Legionella pneumophila* serogroup 1 antigen in the indirect immunofluorescence assay for legionellosis, *J. Clin. Microbiol.*, 16, 979, 1982.
10. **Farshy, C. E., Klein, G. C., and Feeley, J. C.,** Detection of antibodies to Legionnaires disease organism by microagglutination and micro-enzyme-linked immunosorbent assay tests, *J. Clin. Microbiol.*, 7, 327, 1978.
11. **Yonke, C. A., Stiefel, H. E., Wilson, D. L., and Wentworth, B. B.,** Evaluation of an indirect hemagglutination test for *Legionella pneumophila* serogroups 1 to 4, *J. Clin. Microbiol.*, 13, 1040, 1981.
12. **Black, C. M., Pine, L., Reimer, C. B., Benson, R. F., and Wells, T. W.,** Characterization of antibody responses in legionellosis with an immunofluorometric assay, *J. Clin. Microbiol.*, 15, 1077, 1982.
13. **Thompson, T. A. and Wilkinson, H. W.,** Evaluation of a solid-phase immunofluorescence assay for detection of antibodies to *Legionella pneumophila*, *J. Clin. Microbiol.*, 16, 202, 1982.
14. **Kleemola, M. and Kayhty, H.,** Increase in titers of antibodies to *Mycoplasma pneumoniae* in patients with purulent meningitis, *J. Infect. Dis.*, 146, 284, 1982.
15. **Wilkinson, H. W., Cruce, D. D., Fikes, B. J., Yealy, L. P., and Farshy, C. E.,** Indirect immunofluorescence test for Legionnaires' disease, in *Legionnaires: the Disease, the Bacterium, and Methodology*, Jones, G. L. and Hebert, G. A., Eds., Centers for Disease Control, Atlanta, 1979, 111.
16. **Wilkinson, H. W.,** Immune response to *Legionella pneumophila*, in *Manual of Clinical Microbiology*, Rose, N. R. and Friedman, H., Eds. American Society for Microbiology, Washington, D.C., 1980, 500.
17. **Cordes, L. G., Goldman, W. D., Marr, J. S., Friedman, S. M., Band, J. D., Rothchild, E. D., Kravitz, H., Feeley, J. C., and Fraser, D. W.,** Legionnaires' disease in New York City, August-September 1978, *Bull. N.Y. Acad. Med.*, 56, 467, 1980.
18. **Fallon, R. J. and Abraham, W. H.,** Polyvalent heat-killed antigen for the diagnosis of infection with *Legionella pneumophila*, *J. Clin. Pathol.*, 35, 434, 1982.
19. **Goldman, W. D. and Marr, J. S.,** Are air-conditioning maintenance personnel at increased risk of legionellosis?, *Appl. Environ. Microbiol.*, 40, 114, 1980.
20. **Helms, C. M., Renner, E. D., Viner, J. P., Hierholzer, W. J., Wintermeyer, L. A., and Johnson, W.,** Indirect immunofluorescence antibodies to *Legionella pneumophila:* frequency in a rural community, *J. Clin. Microbiol.*, 12, 326, 1980.
21. **Marks, J. S., Tsai, T. F., Martone, W. J., Baron, R. C., Kennicott, J., Holtzhauer, F. J., Baird, I., Fay, D., Feeley, J. C., Mallison, G. F., Fraser, D. W., and Halpin, T. J.,** Nosocomial Legionnaires' disease in Columbus, Ohio, *Ann. Intern. Med.*, 90, 565, 1979.
22. **Snowman, W. R., Holtzhauer, F. J., Halpin, T. J., and Correa-Villasenov, A.,** The role of indoor and outdoor occupations in the seroepidemiology of *Legionella pneumophila*, *J. Infect. Dis.*, 145, 275, 1982.
23. **Storch, G., Hayes, P. S., Hill, D. L., and Baine, W. B.,** Prevalence of antibody to *Legionella pneumophila* in middle-aged and elderly Americans, *J. Infect. Dis.*, 140, 784, 1979.
24. **Macrae, A. D., Appleton, P. N., and Laverick, A.,** Legionnaires' disease in Nottingham, England, *Ann. Intern. Med.*, 90, 580, 1979.

25. **Haley, C. E., Cohen, M. L., Halter, J., and Meyer, R. D.,** Nosocomial Legionnaires' disease: a continuing common-source epidemic at Wadsworth Medical Center, *Ann. Intern. Med.*, 90, 583, 1979.
26. **Dondero, T. J., Jr., Clegg, H. W., II, Tsai, T. F., Weeks, R. M., Duncan, E., Strickler, J., Chapman, C., Mallison, G. F., Politi, B., Potter, M. E., and Shaffner, W.,** Legionnaires' disease in Kingsport, Tennessee, *Ann. Intern. Med.*, 90, 569, 1979.
27. **Broome, C. V., Goings, S. A. J., Thacker, S. B., Vogt, R. L., Beaty, H. N., and Fraser, D. W.,** The Vermont epidemic of Legionnaires' disease, *Ann. Intern. Med.*, 90, 573, 1979.
28. **Grist, N. R., Reid, D., and Najera, R.,** Legionnaires' disease and the traveller, *Ann. Intern. Med.*, 90, 563, 1979.
29. **Berdal, B. P., Farshy, C. E., and Feeley, J. C.,** Detection of *Legionella pneumophila* antigen in urine by enzyme-linked immunospecific assay, *J. Clin. Microbiol.*, 9, 575, 1979.
30. **Tilton, R. C.,** Legionnaires' disease antigen detected by enzyme-linked immunosorbent assay, *Ann. Intern. Med.*, 90, 697, 1979.
31. **Sathapatayavongs, B., Kohler, R. B., Wheat, L. J., White, A., Winn, W. C., Jr., Girod, J. C., and Edelstein, P. H.,** Rapid diagnosis of Legionnaires' disease by urinary antigen detection, *Am. J. Med.*, 72, 576, 1982.
32. **Kohler, R. B., Zimmerman, S. E., Wilson, E., Allen, S. D., Edelstein, P. H., Wheat, L. J., and White, A.,** Rapid radioimmunoassay diagnosis of Legionnaires' disease, *Ann. Intern. Med.*, 94, 601, 1981.
33. **Mangiafico, J. A., Hedlund, K. W., and Knott, A. R.,** Rapid and sensitive method for quantitation of *Legionella pneumophila* serogroup 1 antigen from human urine, *J. Clin. Microbiol.*, 13, 843, 1981.
34. **Wilkinson, H. W. and Fikes, B. J.,** Detection of cell-associated or soluble antigens of *Legionella pneumophila* serogroups 1 to 6, *Legionella bozemanii*, *Legionella dumoffii*, *Legionella gormanii* and *Legionella micdadei* by staphylococcal coagglutination tests, *J. Clin. Microbiol.*, 14, 322, 1981.
35. **Tang, P. W., de Savigny, D., and Toma, S.,** Detection of *Legionella* antigenuria by reverse passive agglutination, *J. Clin. Microbiol.*, 15, 998, 1982.
36. **Wilkinson, H. W.,** unpublished observations.
37. **England, A. C., III,** personal communication.
38. **Wilkinson, H. W.,** unpublished observations.
39. **Wilkinson, H. W.,** unpublished observations.
40. **Wilkinson, H. W.,** unpublished data.
41. **Sampson, J. S., Wilkinson, H. W., Tsang, V.C.W., and Brake, B. J.,** Kinetic- dependent enzyme-linked immunosorbent assay for detection of antibodies to *Legionella pneumophila*, *J. Clin. Microbiol.*, 18, 1340, 1983.
42. **McKinney, R.,** personal communication.

Chapter 3

DIAGNOSIS: DIRECT FLUORESCENT ANTIBODY TEST

Roger M. McKinney

TABLE OF CONTENTS

I. SEROGROUPS AND SPECIES TESTED FOR

The first application of direct fluorescent antibody (DFA) methods for detecting legionellae in clinical specimens was reported by Cherry et al.[1] They prepared conjugates or fluorescein-labeled antibodies for five strains of *Legionella pneumophila*. A conjugate for the Knoxville 1 strain was superior to conjugates for the other four strains and satisfactorily stained all of the strains of *L. pneumophila* that were available at that time. Evidence of serologic variation or multiple serogroups within the *L. pneumophila* species followed rather shortly,[2-8] and DFA reagents were produced to test for these. Further, as new species of *Legionella* were identified, which were shown to be etiologic agents of human pneumonia, appropriate DFA reagents were produced. Presently, DFA testing for all of the recognized serogroups and species of *Legionella* that are known human pathogens necessitates the use of individual conjugates for *L. pneumophila* serogroups 1 to 7, *L. bozemanii, L. micdadei, L. dumoffii,* and *L. longbeachae* serogroups 1 and 2. In addition, *L. gormanii* has been implicated as a human pathogen by serologic evidence, and is also routinely tested for in our laboratory by DFA staining. Other *Legionella* species have been isolated from environmental samples, but they have not been clearly implicated as etiologic agents of human pneumonia. No completely satisfactory animal model or cell culture system has been described to determine whether the *Legionella* species that have been isolated only from the environment are likely etiologic agents of pneumonia in humans. In our laboratory, *Legionella* species that have been found only in the environment are not routinely tested for by DFA staining unless there is serologic evidence that they are implicated in human pneumonia. Since all unfixed clinical specimens are cultured for *Legionella*-like organisms, it appears reasonable that if *Legionella* species other than *L. pneumophila* which have previously been found only in the environment are human pathogens, they will eventually be isolated from clinical specimens through culturing procedures. This practice of testing by DFA only for the proven human pathogens is necessitated by the large number of tests that must be performed to screen for all of the known serogroups and species of *Legionella*. However, experience in our diagnostic laboratory indicates that new *L. pneumophila* serogroups discovered in the environment are likely to be found eventually in human clinical specimens. Therefore, conjugates are included in the testing scheme to detect all known serogroups of *L. pneumophila*. *L. pneumophila* serogroups 3, 5, and 7 are examples where the original isolates were from environmental samples and isolates were obtained later from clinical materials.

II. PREPARATION OF ANTISERA AND CONJUGATES

A. Immunogens and Immunization Schedules

Generally, antisera have been prepared in rabbits by using formalin-killed whole cells as the immunogen. In our laboratory, an initial or priming dose of 4×10^9 cells distributed intradermally in 20 to 40 sites along the back, followed after 1 month of rest by a booster dose given intramuscularly and then weekly by i.v. booster doses for 2 or 3 successive weeks have given satisfactory results with most antigens.[9] However, with some species, i.e., *L. micdadei*, rabbits did not tolerate immunization by the i.v. route with the customary vaccine dosage, but i.m. boosting only was satisfactory for producing high-titered antisera. With some other species or serogroups rabbits did tolerate the i.v. boosting, but antibody response was poor. In such cases better response was obtained by using equal volumes of whole cell vaccine and complete Freund's adjuvant for the priming procedure. Final i.v. boosting appeared always to be superior to i.m. when this route was tolerated by the rabbits.

Wong et al.[10] prepared serogroup-specific antisera for *L. pneumophila* serogroup 1 by immunizing goats with purified antigen that was extracted from cells of a serogroup 1 strain. This appears to be the only report of the use of animals other than rabbits for producing antibodies to *L. pneumophila* for DFA applications. However, with the current rapid development of monoclonal antibody technology, it is likely that mouse antibodies will soon find application in the detection and characterization of legionellae.

B. Isolation of Antibodies from Serum

It is desirable to quantitatively isolate the *Legionella* specific antibodies from the natural antibodies and all other serum components before labeling the antibody preparation with fluorescein isothiocyanate (FITC). This procedure should result in minimizing nonspecific staining or the staining of specimen material other than *Legionella* antigens. Thus, maximum contrast should be provided between the fluorescent-antibody-stained antigens and tissue or other specimen components. In the absence of practical methods specifically and quantitatively to isolate these antibodies, a satisfactory approach is to isolate the gamma globulin fraction, which consists of all *Legionella* specific antibodies in the serum plus all of the other natural gamma globulins that are present. This can be readily accomplished by precipitating with the appropriate concentration of ammonium sulfate.[11] This procedure is practical, requires minimal equipment, and can be performed in any laboratory. However, a considerable amount of time is involved since several precipitations are desirable, and ammonium ion must be completely removed before the gamma globulin is labeled with FITC. An alternative method of isolating gamma globulin is column affinity chromatography with *Staphylococcus* protein A (SPA) covalently bound to sepharose CL4B (Pharmacia Fine Chemicals, Inc., Piscataway, N.J.) as described by Goding.[12] SPA binds to the Fc fragment of the IgG molecule. The affinity of SPA for IgG varies with the subclass as well as with the species source of IgG.[13] The affinity of SPA for rabbit IgG is high, and essentially, IgG can be quantitatively removed from rabbit serum by SPA column affinity chromatography. The IgG is bound at pH 8.0 and released from SPA with 0.1 M acetic acid in 0.85% NaCl. In consideration of the minimal time and labor involved and the high yield and purity of IgG obtained, we chose SPA column affinity chromatography as the method for isolating rabbit antibodies for the preparation of *Legionella* conjugates. IgM and IgA antibodies are essentially lost by this procedure, but these do not appear to be appreciable components of the *Legionella* antibodies obtained by the described immunization schedules. Fraction-collection apparatus and cold room facilities are desirable for SPA column chromatography.

C. Preparation of Conjugates

Conjugates for DFA staining of *Legionella* are prepared by conventional procedures.[9,11] Our practice is to use a relatively high fluorescein to protein (F:P) ratio in the reaction mixture so as to produce conjugates with F:P ratios within the range of 15 to 25 μg of covalently bound fluorescein per milligram of IgG. Rhodamine-labeled whole serum from unimmunized rabbits is used as the diluent.[9] The rhodamine-labeled serum is taken up nonspecifically by tissues and imparts a light orange background, thus providing better contrast and visualization of the specifically stained legionellae. Polyvalent conjugates are prepared by combining monovalent conjugates in appropriate ratios, so as to give bright staining of all serogroups or species that react with the individual or monovalent conjugates.[14] Conjugates that give bright staining at high dilutions are most desirable for preparing polyvalent pools, since nonspecific background fluorescence becomes a serious problem when high concentrations of fluorescein are used. In our experience four or five monovalent conjugates have been the maximum that could be combined to make an effective polyvalent pool.

III. DFA STAINING

In our laboratory, culturing is attempted with all unfixed specimens. Details of the DFA staining procedure are found in other publications.[1,9] Because of the possibility of contamination of water by legionellae, it is advisable to filter-sterilize all water and aqueous solutions used in the staining procedure. If sufficient specimen material is available, eight slide preparations are made for DFA testing. To promote adherence to the slide, with specimens other than mounted tissue sections, the specimens are heat fixed by gentle flaming. As an added safety precaution, where the possibility of the presence of live organism exists (as in smears of respiratory tract fluids and fresh frozen tissue), the previously heat-fixed smear is overlaid with 10% buffered formalin for 10 min. The slide is dipped in distilled water to remove the formalin and the smear is air-dried and stained with DFA conjugate. The staining procedure consists of overlaying the smear with the working dilution of conjugate and incubating in a moist atmosphere for 20 min. Unbound conjugate is removed by rinsing first with a small stream of phosphate-buffered saline (PBS), pH 7.6, being careful to direct the stream to the area above the smear so as to minimize the loss of particulate material from the slide. The slide is then immersed in PBS for 5 min, dipped in distilled water to remove PBS, and air-dried. The air-dried smear is mounted with buffered glycerol, pH 9.0, and covered with a glass coverslip. Microscopic examination is performed with a fluorescence microscope utilizing an appropriate filter system. In our laboratory, routine diagnostic fluorescence microscopy is performed with a Leitz Dialux 20 microscope equipped with a monocular headpiece, 100-W tungsten-halogen lamp, Ploem incident light fluorescence illuminator, and appropriate FITC filter system.

Preferably, positive controls for DFA staining should be culture suspensions and positive preparations of tissue scrapings, sections, or smears of respiratory tract fluids, depending on the types of clinical specimens that are being examined. However, these positive control specimens may not be readily available and one is forced to settle for materials on hand. In our laboratory, conjugates are tested on a weekly basis at working dilution with smears of culture suspensions of the homologous *Legionella* serogroups. New lots of conjugates are also tested with freshly cut and deparaffinized tissue sections that are positive for *Legionella* of the homologous serogroup. This type of specimen provides a stringent test for nonspecific staining. Conjugates that give adequate brightness of specific staining with low background fluorescence on tissue sections should be satisfactory for all types of clinical specimens that are routinely examined for legionellae. When positive DFA results are observed with a particular conjugate and clinical specimen, the appropriate negative controls are provided by conjugates of the heterologous serogroups. DFA testing is performed with each of four polyvalent conjugates which contain antibodies for all of the known serologic variants of legionellae. If positive staining is observed with one polyvalent pool, then further testing is performed with monovalent conjugates, and identification of species and serogroups is made.

A. Frozen or Fresh Tissue

If sufficient tissue is available, a homogenate is prepared by grinding approximately 1 g of tissue with the use of a sterile mortar and pestle, a small mount of alundum powder, and 9 mℓ of water. A sample of the suspended homogenate is used for culturing attempts and for DFA testing. The remaining homogenate is stored at −70°C for possible future reference. If only a small amount of material is available, as with needle biopsies, paraffin embedding and sectioning may be necessary to obtain a satisfactory specimen for DFA testing.

B. Respiratory Tract Fluids

Expectorated sputa, transtracheal aspirates, bronchial washings, pleural fluids, and other specimens from the lower respiratory tract are all satisfactory materials for testing. Culturing attempts are made with respiratory tract fluids by dilution streaking on buffered charcoal yeast extract agar (BCYE)[15] plates. Some experience is necessary to prepare sputum smears of optimum thickness for DFA testing. A smear that is too thin limits the opportunity of finding organisms, whereas nonspecific staining in heavy smears may give so much background fluorescence as to make detection of specifically stained legionellae difficult. It is important that expectorated or aspirated sputum with minimum dilution by saliva be obtained.

C. Formalinized Tissues

Pieces of formalinized tissues are cross-sectioned to present a fresh clean surface. A thin paste is formed by scraping the surface with a scalpel and smears are prepared from this paste. Although some of the tissue falls off of the slide during the staining and washing process, sufficient infected tissue and freed *Legionella* cells adhere to the slide to provide a representative sample of the specimen. Many of the *Legionella* cells may be released from macrophages by the scraping process.

D. Paraffin-Embedded Tissue Sections

Antigens of legionellae organisms in tissue blocks embedded in paraffin will apparently remain intact indefinitely. However, it is preferable to examine tissue sections by DFA on the same day in which the sections are cut and mounted, since the antigens of legionellae in thin mounted sections deteriorate with time, even if removal of paraffin is delayed until shortly before staining. There is some advantage in tissue sections over scrapings, homogenates, or touch imprints in that the tissue morphology is preserved in sections. In the DFA test, the presence of macrophages packed with brightly fluorescent bacteria within the alveoli is highly indicative of legionellosis.

IV. INTERPRETATION OF DFA STAINING RESULTS

The following criteria have been established in our laboratory for DFA testing, interpreting, and reporting results.[1,9] In the absence of readily identifiable fluorescent bacteria, the smear is searched for 5 min before it is reported as DFA negative with a particular conjugate. Smears of formalinized tissue or tissue sections are reported as DFA positive when 25 or more fluorescing cells of the correct morphology are found. If less than 25 fluorescing cells are found on the smear, then the number only is reported. Respiratory fluids from patients with legionellosis frequently contain very small numbers of legionellae so five or more fluorescing cells per smear are reported as positive. If less than five organisms per smear are seen the numbers only are reported. The rationale for this practice is based on experience with many specimens from various sources. Contamination of lung or bronchial tissue by handling procedures must always be considered as a possibility, and this is especially true where very small numbers of cells are found or where all cells appear to be free rather than intracellular in macrophages.

Figure 1 is a photomicrograph of a lung tissue section from a fatal case of legionellosis due to *L. bozemanii* infection.[16] The alveoli, which are in the lower left and right areas of the figure, are filled with macrophages that are packed with *L. bozemanii* cells. Individual free bacteria found on and around the septa were probably spread to these areas by the cutting process. This picture is typical of a very heavy infection with any *Legionella* species as revealed by DFA staining. The preservation of tissue morphology in tissue sections provides a distinct advantage in interpretation of the test. Figure 2 is a photomicrograph of scrapings of formalin-fixed lung tissue from a rare case of a simultaneous infection with two serogroups of *L. pneumophila*. The mixed infection was revealed by double staining with a combination of a rhodamine conjugate for *L. pneumophila* serogroup 1 and a fluorescein conjugate for *L. pneumophila* serogroup 5. The brighter, larger-appearing cells in the figure are serogroup 1 cells that were stained by the rhodamine conjugate. The larger appearance of the serogroup 1 cells is due to an artifact of the photographing procedure. It is remarkable that no mixing of serogroups is observed within individual macrophages that are packed with *Legionella* cells. It would be difficult to demonstrate this type of mixed infection in formalin-fixed tissue by means other than DFA staining.

Heavy *Legionella* infection is easily detected by DFA staining. DFA staining results may be obtained within 2 hr after receipt of the specimen. In the case of biopsy or respiratory

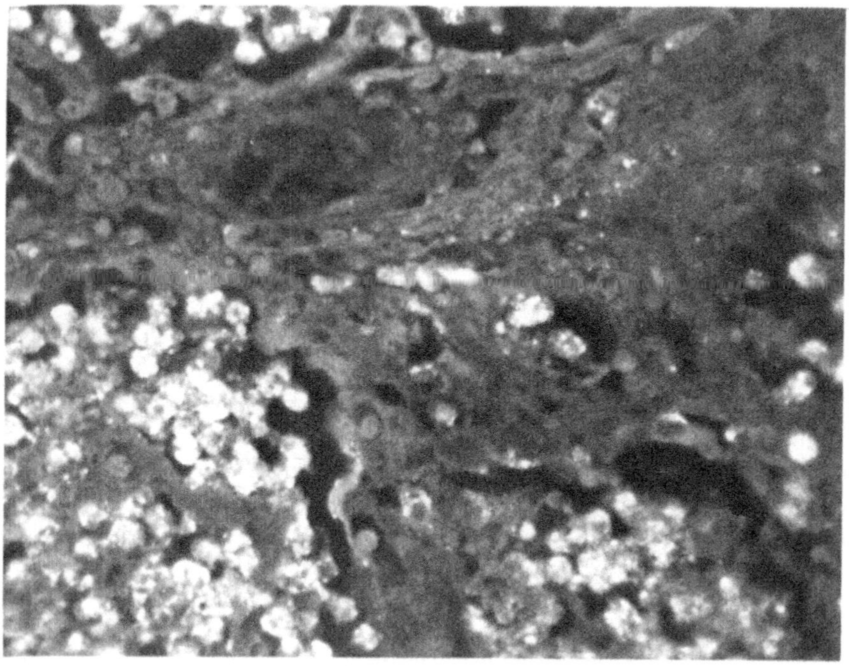

FIGURE 1. Photomicrograph of a lung tissue section positive for *L. bozemanii* by DFA staining. Note the alveoli in the lower left and lower right are filled with macrophages which are packed with DFA-stained *L. bozemanii* organisms. Fluorescent free and intracellular *L. bozemanii* organisms seen on the septum may have been spread to this area as a result of the cutting process. (From Thomason, B. M. et al., *Ann. Intern. Med.*, 91, 674, 1979. With permission.)

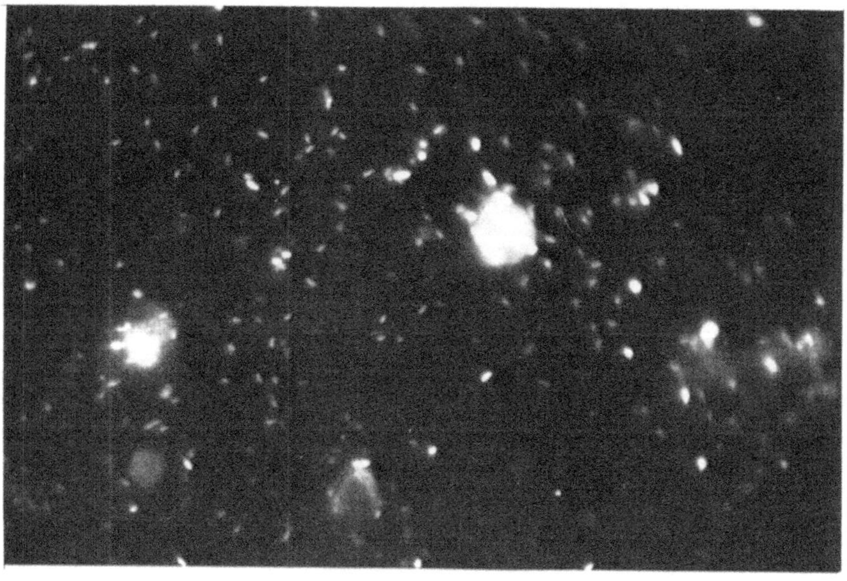

FIGURE 2. Photomicrograph of a lung tissue scraping that is DFA$^+$ for *L. pneumophila* serogroups 1 and 5. Double staining was performed with a mixture of rhodamine-labeled specific antibody for *L. pneumophila* serogroup 1 and fluorescein-labeled specific antibody for *L. pneumophila* serogroup 5. The brighter and larger-appearing organisms are the rhodamine-stained serogroup 1 cells. No mixing of serogroups was observed within any of the infected macrophages. The photomicrograph was obtained by a double exposure using first the filter system that allows rhodamine observation and then the filter system that allows fluorescein observation.

Table 1
HUMAN RESPIRATORY SYSTEM SPECIMENS TESTED FOR *LEGIONELLA PNEUMOPHILA*[a]

Type of specimens	Total no. specimens tested[b]	Total no. DFA +[b]	No. DFA − (fresh or frozen)	No. DFA + (fresh or frozen)	No. *L. pneumophila* isolates
Sputa, bronchial washings and transtracheal aspirates	463	15	256	12	4
Lung tissue and pleural fluids	1488	212	571	52	27

[a] Tests conducted in one laboratory at the CDC by DFA staining during the period from September 1, 1977 to December 31, 1979.
[b] Includes formalinized and fresh or frozen specimens.

From Blackmon, J. A. et al., *Am. J. Pathol.*, 103, 453, 1981. With permission.

fluid specimens it is obvious that a positive DFA test is of great value in providing an early diagnosis and in guiding the clinician in the choice of proper therapy. However, negative DFA results should be interpreted with caution. *Legionella* organisms are frequently not found in improperly collected sputum that is highly diluted with saliva, or in very dilute bronchial wash specimens. Lung tissue specimens may be taken from a noninfected area of the lung or from an area which was once infected but from which *Legionella* cells are no longer present due to tissue breakdown or to previous antibiotic therapy. Thus, a negative DFA result by itself does not rule out the possibility of legionellosis and must be taken into consideration along with all other clinical information. With fresh or frozen specimens, where culturing is possible, DFA testing has been about twice as sensitive as culturing in our laboratory. In only one instance during a 3-year period of testing was a tissue specimen reported as DFA negative and later found to be culture positive. Further and more intensive examination by DFA in this case revealed that the specimen was also DFA positive but with very few organisms present. Results of DFA testing and culturing in our laboratory over an approximate 2-year period are shown in Table 1.[17] The small number of isolates obtained from sputum specimens relative to the number of DFA positives may be due to small numbers of *L. pneumophila* and to problems of heavy overgrowth of normal flora in sputum specimens. The use of selective growth media is desirable for this type of specimen. In contrast to our results, Edelstein et al.[18] found culturing of specimens to be more sensitive than DFA testing. In agreement with our laboratory practice, these authors concluded that both DFA testing and culturing of appropriate specimens should be performed as well as IFA testing of sera to achieve maximum sensitivity and specificity in diagnosing legionellosis.

The number of clinical specimens found positive for *Legionella* by DFA testing in our laboratory during 1979 are listed by *L. pneumophila* serogroups and other *Legionella* species in Table 2.[17] *L. pneumophila* far outweighed the other *Legionella* species as a cause of legionellosis, and approximately 75% of all *L. pneumophila* infections detected were due to serogroup 1.

V. SPECIFICITY

Some serogroup cross-reactivity is observed in DFA testing with conjugates for the seven *L. pneumophila* serogroups. Serogroup 2 and serogroup 3 conjugates and cells show reciprocal cross-reactivity of low intensity which disappears when the conjugates are diluted to the working concentration for the respective serogroup. Conjugates for serogroup 6 give

Table 2
RESPIRATORY SYSTEM SPECIMENS FOUND POSITIVE FOR *LEGIONELLA* BY DIRECT FLUORESCENT ANTIBODY (DFA) TESTING[a]

	L. pneumophila serogroup						*L. boze-manii*	*L. micdadei*	*L. dumoffii*
	1	2	3	4	5	6			
No. DFA +	44	3	4	4	0	2	1	1	1

[a] DFA conjugates were not available for *L. pneumophila* serogroups 5 and 6, *L. bozemanii*, *L. micdadei*, and *L. dumoffii* organisms for the entire year. However, most of the DFA-negative lung tissue specimens were stored and reexamined with these conjugates as they became available. Tests conducted at the CDC in 1979; distribution by *L. pneumophila* serogroup and other *Legionella* species.

From Blackmon, J. A. et al., *Am. J. Pathol.*, 103, 453, 1981. With permission.

irregular spotted staining of cells of serogroups 1, 2, and 4. There is strong reciprocal cross-reactivity between serogroup 3 and serogroup 6 conjugates and cells. This cross-reactivity is practically eliminated by diluting the conjugate to working concentration for its respective serogroup, and can be completely removed by absorbing the conjugate or antiserum with formalin-killed cells of the heterologous serogroup. Thus, serogroups 3 and 6 possess a rather strong common antigen plus a serogroup-specific antigen, all of which are active in DFA staining. Currently, there are only two isolates of serogroup 7, the first of which was obtained from a showerhead and the second from an autopsy lung specimen.[8] No cross-reactivity has been observed between serogroup 7 and other *L. pneumophila* serogroups. Serogroup-specific antigens of *L. pneumophila* were demonstrated by immunoelectrophoresis with hot saline extracts of a representative strain of each of the serogroups 1 to 4.[4] More recently, Joly and Kenny demonstrated the presence of a serogroup-specific antigen for representative strains of each of the serogroups 1 to 4 by the more elegant crossed-immunoelectrophoresis technique.[19] In addition, Joly and Kenny demonstrated the presence of 27 antigens of *L. pneumophila* that are common to representative strains of serogroups 1 to 6. Obviously, the serogroup-specific antigens are major surface antigens. The other common antigens that are obtained by the extraction procedures are either nonsurface or may be sterically hindered from being revealed by the usual DFA staining. Enzymatic degradation of outer cell membranes may explain occasional observations of *L. pneumophila* cells in clinical specimens giving irregular positive DFA reactions with conjugates for more than one serogroup, whereas the isolate that is obtained from the specimen reacts with only one serogroup-specific conjugate. *L. pneumophila* serogroup 1 is a serologically diverse group within itself, as has been shown in our laboratory by absorbing conjugates for various serogroup 1 strains with formalin-killed cells of other serogroup 1 strains. Subclassification schemes within serogroup 1 can be devised on this basis. Serologic diversity within serogroup 1 was further demonstrated in our laboratory with monoclonal antibodies produced for two serogroup 1 strains. Although there is one surface antigen common to *L. pneumophila* serogroup 1 strains, a number of other surface antigens can be demonstrated to be present in various subgroups of serogroup 1 that are absent in all other *L. pneumophila* serogroups. This serologic diversity within serogroup may provide useful markers for epidemiologic studies.

Cross-reactivity within the *Legionella* genus has not been observed in our laboratory with conjugates for *L. micdadei*, *L. dumoffii*, or *L. gormanii*. Conjugates for *L. longbeachae* serogroups 1 and 2 show a weak reciprocal cross-reactivity at low dilution, but they are serogroup specific at the respective working concentrations. *L. bozemanii* and *L. jordanis* conjugates and cells show strong reciprocal cross-reactivity, but this can be removed by absorbing with cells of the heterologous species.

DFA staining of flagella on *L. pneumophila* strains was first reported by Thomason et al.[20] They first observed DFA staining of polar flagella on *L. pneumophila* strains Bellingham 1 and Bloomington 2. Further examination of *L. pneumophila* strains with Clark's modification of Leifson's flagella stain revealed flagella on 33 of 41 strains.[20] Hebert et al.[21] found that conjugates of antisera for *L. bozemanii* (strain WIGA). *L. micdadei* (strains TATLOCK and HEBA), and *L. pneumophila* (polyvalent serogroups 1 to 4), when used at a high concentration relative to the routine use dilution for cellular staining, gave indiscriminate staining of flagella of strains of all three species. The flagella of representative strains of *L. pneumophila* serogroups 1, 2, and 3 have been shown to be antigenically similar and to form a single line of identity in immunodiffusion with antisera to flagella of each of the three serogroups.[22] No complete systematic study to determine the extent to which the common identify of flagellar antigens observed in *L. pneumophila* serogroups 1 to 3 extends throughout the other *Legionella* serogroups has been reported. Further, it is not known to what extent, if any, this commonality of flagellar antigens extends to other Gram-negative bacteria outside the genus *Legionella*. Studies of flagellar antigens among legionellae have been impeded to some extent by the tendency of most strains to lose their flagella after several passages on CYE agar. DFA staining of flagella appears to be of little diagnostic value because of the cross-reactivity observed in flagellar staining and also because flagellated legionellae are rarely observed in clinical specimens. However, flagellated legionellae are frequently seen in isolates obtained from infected yolk sacs or from peritoneal swabs from infected guinea pigs. Awareness of the cross-staining of flagella is important in these cases as, occasionally with heterologous staining systems, they appear as masses of brightly fluorescent flagella in the absence of any visible cell bodies. In these cases it is obvious that organisms are present, but antibody in the conjugate does not recognize surface cellular antigens of the organism and the organism may represent a previously unrecognized *Legionella* species or serogroup.

A limited number of cross-reactions of *Legionella* conjugates with strains of bacterial species outside of the genus *Legionella* have been noted. Some strains of mycobacteria give minor cross-staining with *L. pneumophila* serogroup 1 conjugates.[23] However, when Freund's complete adjuvant is not used with immunogen this cross-reactivity is minimal and should not present a problem in identifying legionellae.[24] Cross-reactivity of *L. pneumophila* serogroup 1 conjugate (Knoxville 1 strain) has been noted with three strains of *Pseudomonas fluorescens*,[1,25] three strains of *Bacteroides fragilis*,[26] one strain of *Pseudomonas alcaligenes*,[25] and one strain of *Flavobacterium xanthomonas*.[25] One strain of *Pseudomonas alcaligenes* was found to react with *L. pneumophila* serogroup 3 conjugate (strain Bloomington 2), and one strain of *Pseudomonas maltophilia* reacted with *L. micdadei* conjugate (strain TATLOCK).[25] It should be emphasized that these cross-reactions are observed only with one strain or with a very limited number of the strains within each species that were tested. Cross-reactivity can be removed in all of these cases by absorbing conjugates with cells of the appropriate organisms, and the absorbed conjugates retain sufficient specific antibody to satisfactorily stain members of the homologous *Legionella* serogroup. However, the number of isolates or strains of heterologous organisms that do cross-react with the *Legionella* conjugates is very limited, and the practicality of absorbing conjugates with the hope of obtaining absolute specificity is questionable. Cross-reactivity due to the presence of antigens in common with more than one bacterial species is a problem with all diagnostic tests based on immunologic reactions. It is remarkable that the prevalence of cross-reactivity of *Legionella* conjugates with bacterial species outside of the genus *Legionella* is so low. The only absolute method of identifying a *Legionella* infection is to isolate the bacterium from clinical specimens under conditions which exclude the possibility of specimen contamination. However, culturing and characterizing the organism is a relatively slow procedure which only rarely benefits the patient. DFA staining, when properly interpreted, is a highly reliable and valued test in the diagnosis of legionellosis.

DISCLAIMER

Use of trade names is for identification only and does not imply endorsement by the Public Health Service or by the U.S. Department of Health and Human Services.

REFERENCES

1. **Cherry, W. B., Pittman, B., Harris, P. P., Hebert, G. A., Thomason, B. M., Thacker, L., and Weaver, R. E.**, Detection of Legionnaires disease bacteria by direct immunofluorescent staining, *J. Clin. Microbiol.*, 8, 329, 1978.
2. **McKinney, R. M., Thomason, B. M., Harris, P. P., Thacker, L., Lewallen, K. R., Wilkinson, H. W., Hébert, G. A., and Moss, C. W.**, Recognition of a new serogroup of Legionnaires disease bacterium, *J. Clin. Microbiol.*, 9, 103, 1979.
3. **Edelstein, P. H., Meyer, R. D., and Finegold, S. M.**, Isolation of a new serogroup of Legionnaires' disease bacterium, *Lancet*, 2, 1172, 1978.
4. **McKinney, R. M., Thacker, L., Harris, P. P., Lewallen, K. R., Hébert, G. A., Edelstein, P. H., and Thomason, B. M.**, Four serogroups of Legionnaires' disease bacteria defined by direct immunofluorescence, *Ann. Intern. Med.*, 90, 621, 1979.
5. **Nagington, J., Wreghitt, T. G., and Smith, D. J.**, How many Legionnaires?, *Lancet*, 2, 536, 1979.
6. **England, A. C., III, McKinney, R. M., Skaliy, P., and Gorman, G. W.**, A fifth serogroup of *Legionella pneumophila*, *Ann. Intern. Med.*, 93, 58, 1980.
7. **McKinney, R. M., Wilkinson, H. W., Sommers, H. M., Fikes, B. J., Sasseville, K. R., Yungbluth, M. M., and Wolf, J. S.**, *Legionella pneumophila* serogroup six. Isolation from cases of legionellosis, identification by immunofluorescence staining, and immunologic response to infection, *J. Clin. Microbiol.*, 12, 395, 1980.
8. **Bibb, W. F., Arnow, P. M., Dellinger, D. L., and Perryman, S. R.**, Isolation and characterization of a seventh serogroup of *Legionella pneumophila*, *J. Clin. Microbiol.*, 17, 346, 1983.
9. **Cherry, W. B. and McKinney, R. M.**, Detection of Legionnaires' disease in clinical specimens by direct immunofluorescence, in *"Legionnaires" the Disease, the Bacterium and Methodology*, Jones, G. L. and Hébert, G. A., Eds., Centers for Disease Control, Atlanta, 1979, 91.
10. **Wong, K. H., Schalla, W. O., Arko, R. J., Bullard, J. C., and Feeley, J. C.**, Immunochemical, serologic, and immunologic properties of major antigens isolated from the Legionnaires' disease bacterium, *Ann. Intern. Med.*, 90, 634, 1979.
11. **Hébert, G. A., Pittman, B., McKinney, R. M., and Cherry, W. B.**, The Preparation and Physicochemical Characterization of Fluorescent Antibody Reagents, Centers for Disease Control, Atlanta, 1972.
12. **Goding, J. W.**, Conjugation of antibodies with fluorochromes: modifications to the standard methods, *J. Immunol. Methods*, 13, 215, 1976.
13. **Goding, J. W.**, Use of *Staphylococcal* protein A as an immunological reagent, *J. Immunol. Methods*, 20, 241, 1978.
14. **Thomason, B. M., Harris, P. P., Lewallen, K. R., and McKinney, R. M.**, Preparation and testing of a polyvalent conjugate for direct fluorescent-antibody detection of *Legionella pneumophila*, *Curr. Microbiol.*, 2, 357, 1979.
15. **Pasculle, A. W., Feeley, J. C., Gibson, R. J., Cordes, L. G., Myerowitz, R. L., Patton, C. M., Gorman, G. W., Carmack, C. L., Ezzell, J. W., and Dowling, N. J.**, Pittsburgh pneumonmia agent: direct isolation from human lung tissue, *J. Infect. Dis.*, 141, 727, 1980.
16. **Thomason, B. M., Harris, P. P., Hicklin, M. D., Blackmon, J. A., Moss, C. W., and Matthews, F.**, A *Legionella*-like bacterium related to WIGA in a fatal case of pneumonia, *Ann. Intern. Med.*, 91, 673, 1979.
17. **Blackmon, J. A., Chandler, F. W., Cherry, W. B., England, A. C., III, Feeley, J. C., Hicklin, M. D., McKinney, R. M., and Wilkinson, H. W.**, Legionellosis review article, *Am. J. Pathol.*, 103, 453, 1981.
18. **Edelstein, P. H., Meyer, R. D., and Finegold, S. M.**, Laboratory diagnosis of Legionnaires' disease, *Am. Rev. Resp. Dis.*, 121, 317, 1980.
19. **Joly, J. R. and Kenny, G. E.**, Antigenic analysis of *Legionella pneumophila* and *Tatlockia micdadei (Legionella micdadei)* by two-dimensional (crossed) immunoelectrophoresis, *Infect. Immun.*, 35, 721, 1982.
20. **Thomason, B. M., Chandler, F. W., and Hollis, D. G.**, Flagella on Legionnaries' disease bacteria: an interim report, *Ann. Intern. Med.*, 91, 224, 1979.

21. **Hébert, G. A., Moss, C. W., McDougal, L. K., Bozeman, F. M., McKinney, R. M., and Brenner, D. J.,** The rickettsia-like organisms TATLOCK (1943) and HEBA (1959): bacteria phenotypically similar to but genetically distinct from *Legionella pneumophila* and the WIGA bacterium, *Ann. Intern. Med.*, 92, 45, 1980.
22. **Elliott, J. A. and Johnson, W.,** Immunological and biochemical relationships among flagella isolated from *Legionella pneumophila* serogroups 1, 2, and 3, *Infect. Immun.*, 33, 602, 1981.
23. **Katz, S. M. and Hammel, J.,** Confusion of *Legionella pneumophila* and *Mycobacterium tuberculosis*, *N. Engl. J. Med.*, 300, 1490, 1979.
24. **Cherry, W. B. and McKinney, R. M.,** *Legionella*, mycobacteria, and Freund's adjuvant, *N. Engl. J. Med.*, 301, 1242, 1979.
25. **Orrison, L. H., Bibb, W. F., Cherry, W. B., and Thacker, L.,** Determination of antigenic relationships amand non-legionellae by direct fluorescent antibody and immunodiffusion tests, *J. Clin. Microbiol.*, 17, 332, 1983.
26. **Edelstein, P. H., McKinney, R. M., Meyer, R. D., Edelstein, M. A. C., Krause, C. J., and Finegold, S. M.,** Immunologic diagnosis of Legionnaires' disease: cross-reactions with anaerobic and microaerophilic organisms and infections caused by them, *J. Infect. Dis.*, 141, 652, 1980.
27. **Bibb, W. F., et al.,** manuscript in preparation.
28. **McKinney, R. M.,** manuscript in preparation.

Chapter 4

DIAGNOSIS: NEW CONCEPTS

Richard W. Gilpin

TABLE OF CONTENTS

INTRODUCTION

One exciting aspect of the laboratory diagnosis of Legionnaires' disease is the continuing search for better laboratory tests. The goal is to provide accurate, confirmatory information to physicians in the shortest amount of time with the least expense in terms of personnel time and cost per test.

The preceding chapters written by experts in the field covered methods currently used for diagnosis of legionellosis. Some of the new concepts will be reviewed in this chapter. This will include modifications of accepted tests as well as those which are of more recent origin. The chapter is divided into two sections: (1) tests that are used to detect a patient's immunological response to *Legionella* disease and (2) those designed to identify *Legionella* antigens or intact bacteria in patient specimens. The procedures are classified further by the types of equipment needed to perform the tests.

II. TESTS FOR ANTIBODY

A. Fluorescence Microscopy

Heat-killed instead of formalin-killed *Legionella* antigen seems to work best in the indirect immunofluorescence assay (IFA) that was described in Chapter 3.[1] Patient sera are usually diluted in phosphate buffered saline (PBS). One modification was the use of an extract from *Escherichia coli* strain 013:K92:H4 for dilution of patient sera to remove nonspecific antibody reactions. Acute-phase titers, although lower, made it possible to detect seroconversions because convalescent titers were not reduced when *L. pneumophila* serogroup 1 antigen was used.[2] However, when polyvalent serogroup antigen was substituted, a loss of sensitivity was found. Therefore, the use of this immunosorbent may not be necessary in most situations. Comparisons between *Mycoplasma pneumoniae*, *Chlamydia psittaci* and *Legionella* spp. antigens with paired patient sera having concurrent titers to *Legionella* and either *Mycoplasma* or *Chlamydia* indicated that the IFA test did not produce nonspecific cross-reactions to either *Mycoplasma* or *Chlamydia*.[3] Also, adsorption of sera with *Legionella* spp. removed only antibody to *Legionella*.[3] The presence of cross-reactions between various species of *Legionella* may occur, however.

Determination of IFA serum titers with sheep anti-human IgM in addition to IgG may permit earlier detection of IgM antibody in patient serum before significant titers of IgG can be detected.[4] Therefore, anti-human IgM would be appropriate to incorporate into routine IFA testing for early presumptive diagnosis of legionellosis. Although the IFA test requires experienced personnel to interpret staining intensities, artifacts, and nonspecific background staining reactions, this technique remains the reference to which other antibody tests are compared.

B. Electron Microscopy

1. Reverse Immunoferritin Test (RIFEM)

Patient antibody was titered by floating carbon-coated grids containing formalin-killed *L. pneumophila* and normal yolk-sac membrane upside down on serial dilutions of patient serum for 30 min at 37°C.[5] The grids were washed and incubated on rabbit anti-human ferritin-conjugated globulin, washed, and examined for the presence of ferritin granules over the entire surface of the known *Legionella*. Comparisons of serum titers from 12 patients by the IFA and RIFEM tests were within one dilution of each other.[5] However, titration of patient sera by RIFEM is time and technique intensive, which makes it more appropriate for research rather than diagnostic applications.

C. Colorimetric/Fluorometric Tests

1. Enzyme-Linked Immunosorbant Assay (ELISA)

This test has been used to detect antibody in patient sera to heat-killed *L. pneumophila* serogroup 1 soluble antigen.[6] Wells in polystyrene micro-ELISA plates coated with Legionella antigen received serial dilutions of patient serum. After incubation for 1 hr at 37°C the wells were washed with buffered saline. An appropriate dilution of horseradish peroxidase-conjugated antihuman globulin was added and incubated for 1 hr at 37°C. The wells were again washed and orthophenylenediamine was added. The plates were incubated for an additional 30 to 60 min at 37°C and the reaction was stopped by adding H_2SO_4. The titer was the highest dilution of patient serum producing a color different from the negative control (no serum). More comparisons with the IFA procedure, which takes less time to perform, will indicate whether this procedure is superior. The advantage of the ELISA method is that results can be determined visually or by colorimetry.

2. Solid-Phase Immunofluorescence Test (FIAX)

This had previously been used for detection of antibody to viral and fungal antigens. The method was tested with immune sera from guinea pigs vaccinated with *Legionella*.[7] Sonicated, formalin-killed *L. pneumophila* were adsorbed to one side of a plastic paddle stick. The reverse side of the paddle contained no antigen (control). Paddles were placed in diluted guinea pig sera and shaken for 30 min at room temperature. The paddles were removed, washed free of unbound serum, and incubated for 40 min with fluorescein isothiocyanate-conjugated rabbit anti-guinea pig IgG. After washing, the fluorescence on the paddles was measured with a special fluorometer. The fluorescence of the control side of each paddle was subtracted from the fluorescence of the side containing adsorbed *Legionella*. The method detected end points with a low coefficient of variation between assays. Although need for a UV light microscope was eliminated, a special fluorometer was used to measure fluorescence of the paddles.

A recent comparison was made between the IFA and FIAX tests using paired human serum from patients with legionellosis.[8] Both tests detected titers to *L. pneumophila* serogroup 1 within one twofold dilution of each other, but the IFA test was significantly more sensitive and used lower quantities of test reagents.

3. Immunofluorometric Assay (IFMA)

An assay originally developed to measure titers of antibody reagents and to screen for monoclonal antibody production was used to detect antibody to *Legionella* in patient sera.[9] *L. pneumophila* were formalin killed, washed, and incubated at 37°C with serial dilutions of patient sera for 1 hr. Suspensions were centrifuged and the resulting pellets were washed. Fluorescein-isothiocyanate-conjugated goat antihuman antibody was added to the tubes and incubated at 37°C for 1 hr. After washing, the fluorescent intensity of each *Legionella* suspension was measured in a fluorometer. Data analysis was performed by computer. The procedure detected antibody levels to *Legionella* in patient sera that were negative by the IFA assay. Although a fluorometer and appropriate computer software were needed, the procedure did not require the level of experience needed to perform the IFA assay. But, as with the FIAX assay, more reagents were needed to perform this assay. Further miniaturization may make these solid-phase immunofluorescent assays more economical.

D. Visible Agglutination or Precipitin Tests

These tests avoid the use of fluorescent or enzyme-conjugated antibodies for detection of antibody to *Legionella* test antigen.

1. Microagglutination Test (MA)

This procedure was applied to detection of IgM antibody to *L. pneumophila* in patient sera.[6] Several authors have used this procedure for serologic surveys because several sera can be screened on one microtiter plate. Heat-killed *Legionella* stained with safranin were added to microtiter wells containing serially diluted patient serum. After incubation at room temperature for 16 hr followed by 2 hr at 4°C, visible agglutination end points were determined to be the highest dilution of serum that did not result in formation of a distinct button at the bottom of a microtiter well.[6] This procedure was used to evaluate the major F-1 antigen of *L. pneumophila*.[10] The MA test did show serogroup specificity against serogroups 1 to 4 *Legionella*.[11] Minor modifications of the MA test involved different serum dilutions and different methods for preparing the *Legionella* antigen.[12]

A modified microagglutination test that could be performed in 30 min instead of the usual overnight incubation was recently reported.[13] *L. pneumophila* were incubated overnight at 37°C in 2% formalin buffered saline (FBS), pH 7.0. An equal volume of phosphate buffered saline (PBS), pH 6.4, containing 0.005% (w/v) safranin was added; the mixture was incubated for 1 hr at room temperature, and centrifuged at 3000 × g for 15 min. The pellet was then suspended in PBS, pH 6.4, for use as antigen. Doubling dilutions of patient serum were placed in microtiter wells, antigen was added, and the plates were sealed and shaken for 10 sec. After 10-min incubation at room temperature, the plates were centrifuged at 300 × g for 4 min, inclined at a 70° angle for 10 min, and scored as negative (tear-drop pattern) or positive (clear button). Tests of patient sera produced a 97% agreement with the IFA test.[13]

The shorter incubation times possible with the modified MA technique justify further evaluation in diagnostic laboratories.

One drawback of the MA tests is that IgM is more reactive than IgG antibody. Therefore, other methods are needed to detect IgG. One advantage of the MA test is that equipment such as an UV light microscope or fluorometer are not needed. One disadvantage is occasional difficulty reading end point titers, but this may be avoided if the antigen is dispersed uniformly.

2. Indirect Hemagglutination Test (IHA)

This procedure was also performed in microtiter plates.[14] Turkey erythrocytes containing absorbed *L. pneumophila* soluble antigens were added to microtiter wells containing patient sera diluted 1:16 and incubated at room temperature for 2 hr. Sera which caused agglutination of the sensitized erythrocytes were then serially diluted and retested to determine their antibody titer to *Legionella*. Turkey erythrocytes without *Legionella* antigen served as the control. Agreement with the IFA test was 95% but IHA titers were somewhat lower. Another study evaluated the IHA test with *L. pneumophila* serogroup 1 to 4 antigens.[15] Agreement with the IFA test was 97% overall, but in many instances one test was positive when the other test was negative.[15] Sera fractionated on sucrose gradients showed that the IHA test detected IgM antibody that would not be picked up by the IFA procedure unless fluorescein-conjugated anti-IgM antibody was used. Occasionally, the IHA test did not detect IgG antibody. For screening of sera this technique appeared to be suitable and slightly more sensitive than the MA test. Another study indicated that screening sera with polyvalent *Legionella* antigen produced lower titers, although agreement with the IFA test remained at 97%.[16]

3. Immune Adherence Hemagglutination Test (IAHA)

Heat-killed *L. pneumophila* test antigen was suspended in PBS, pH 7.4, and stored at −80°C. Patient sera were diluted 1:4, complement was heat inactivated at 56°C, and then serially diluted in microtiter wells.[17] Antigen was added and the plates were incugated at 37°C for 1 hr. Diluted guinea pig complement was added and incubation was continued for 40 min at 37°C. After the addition of ethylenediaminetetraacetate and dithiothreitol, a suspension of human type O erythrocytes (RBC) were added and incubation was continued for 1 hr at room temperature. Complete agglutination of RBC was considered positive for the presence of antigen-antibody-complement immune complexes. Patient sera that were IFA positive were also positive in the test (25 of 26 sera). Although IAHA titers were often lower, they were specific for *Legionella* IgM and IgG antibody in patient sera.

4. Immunodiffusion Test (ID)

Soluble test antigen was obtained from the supernatant fluid after centrifugation of heat-killed suspensions of *L. pneumophila*.[18] Antigen was placed in the center well of a small petri dish containing 1% agar and four wells, 4 mm from the central well, were filled with patient sera to test for antibody to *Legionella* soluble antigen. During incubation at 37°C for 72 hr, the presence of precipitin bands was noted. One to two bands appeared after incubation in many of the positive patient sera. Test results correlated with IgM antibody. Sera with high IgG titers and low titers of IgM antibody may not produce a positive ID test, however.

5. Card Agglutination Titer (CAT)

A card agglutination test was developed to provide a facile method to determine the antibody titer to *Legionella* in patient serum.[19] The rapid plasma reagin card test for syphilis was modified for this purpose. Serogroup 1, *L. pneumophila* grown on charcoal-yeast extract (CYE) agar were suspended in PBS and steamed for 60 min. After washing, the diluted suspension of *Legionella* antigen was mixed with crystal violet (0.01% final concentration). Patient sera (50 μℓ) were serially diluted up to 1:512 with buffered saline directly within each ring on the card. The diluted sera were spread over the rings and crystal-violet-stained *Legionella* were added. The card was rotated by hand for 15 to 20 sec. If no agglutination was observed, the card was placed on a card test rotator for 3 to 4 min at 100 rotations per minute. The end point titer was the highest dilution of patient serum producing agglutination. Although the card that was not as sensitive as the IFA test, the results were specific. Therefore, the method could be used for rapid screening in laboratories not equipped to do fluorescent microscopy.

6. Counterimmunoelectrophoresis (CIE)

Diagnosis of legionellosis by detection of antibody in patient serum was performed by using soluble *Legionella* antigen.[20] Heat-killed *Legionella* were disrupted by sonication and used without removing the large particles. Opposing wells were cut in 1% agarose-covered microscope slides. Diluted patient test serum was placed in a well toward the anode and soluble antigen was placed in the cathode well. Electrophoresis at 10 mA per slide was performed at room temperature for 90 min. After incubation at 4°C for 30 min, the slides were washed and examined for the presence of precipitin bands. Slides were preserved by drying and staining with naphthalene black. As might be expected, the serogroup-specific serum titers by CIE were lower than the IFA titers.

E. Cell-Mediated Immunity Tests

1. Lymphocyte Blastogenic Response

Lymphocytes were isolated from patients with culture-positive or IFA-confirmed legionellosis and those with other bacterial pneumonias. The lymphocytes were cultured in the presence of sonicated, formalin-killed *L. pneumophila* for 6 days. ^{3}H-Thymidine was added to the culture 6 hr before harvest to measure blastogenesis by incorporation of thymidine. Among a small group of patients tested, the assay appeared to detect an immune response to *Legionella* disease before positive titers were detected by IFA.[21,22] More studies would be needed to confirm these results. Although this test is difficult to perform, the ability to make a laboratory diagnosis before the primary humoral response can be detected by IFA would be quite useful.

2. Skin Test

A study of guinea pigs sensitized with *L. pneumophila* indicated that intracutaneous injection of antigen produced a delayed hypersensitivity reaction within 24 to 48 hr.[23] Control animals did not produce this reaction. In another study, *L. pneumophila* serogroup 1 to 4 bacteria, or antigens extracted with lithium acetate, were used to sensitize guinea pigs.[24] Four different procedures were used. Animals sensitized by heat-killed and live *Legionella* produced delayed hypersensitivity reactions within 48 hr when skin tested with cross-reactive antigens. Skin tests of immunized guinea pigs given a challenge dose of live *Legionella* were also positive. Skin tests of normal control animals were negative. Therefore, the diagnostic use of a skin test deserves further study.

F. Conclusions

Some tests to detect the presence of antibody or cellular antibody response are quite sophisticated and require specific equipment and/or training of laboratory personnel. The IFA test remains the baseline to which other serological tests are compared. Although this test does require an experienced observer to interpret end point titers, the IFA seems to be somewhat more sensitive than agglutination and precipitin tests described in this section if fluorescein-conjugated antibody to both human IgG and IgM are used. Many of the other methods require the use of more reagents, thus making each test more expensive.

The ability to make a diagnosis of legionellosis before specific humoral antibody can be detected in patient sera would make serological tests less retrospective. The method described to detect sensitization of lymphocytes or other similar procedures may be of more use to the diagnostician than tests for humoral antibody. The relevance of this and the skin test may become more apparent as further studies are performed.

III. TESTS FOR *LEGIONELLA* ANTIGEN

A. Electron Microscopy

1. Immunoperoxidase-Labeled Antibody

Visualization of surface antigens can be enhanced with specific antibody coupled to an enzyme. Peroxidase-conjugated anti-*Legionella* F-1 antibody was used to locate a high-molecular-weight surface antigen.[25] Saponin-treated suspensions of *L. pneumophila* were fixed with paraformaldehyde and gluteraldehyde.[26] After washing, the cells were preincubated with 5% bovine serum albumin (BSA) followed by peroxidase-conjugated rabbit antibody to F-1 antigens.[25] After incubation for 45 min at room temperature and washing, the stain reaction mixture[26] was added. After 15-min incubation the bacteria were washed, post-fixed in osmium, embedded in epon, sectioned, and observed with an electron microscope. Thin sections showed the surface location of the F-1 antigen on *L. pneumophila*. The staining reaction was serogroup specific. Application of this method to locating *Legionella* in tissue specimens may be of immediate interest to researchers.

The immunoperoxidase method was also used to study the antigens on *Legionella* flagella.[27] Flagellin recovered from agar cultures of *L. pneumophila*, serogroup 1, was used to immunize rabbits. Rabbit antibody to flagellin as well as to F-1 antigen were used for indirect immunoperoxidase assays. Glutaraldehyde-fixed *Legionella* were adsorbed onto carbon-coated grids, washed, and incubated with normal mouse serum for 20 min. After subsequent washing and incubation with normal mouse serum, the grids were incubated for 20 min with the rabbit antiserum. Washed grids were then incubated with horseradish peroxidase-conjugated goat anti-rabbit IgG for the same time period. After washing, diaminobenzidine tetrahydrochloride and hydrogen peroxide were added. The grids were then washed and stained with phosphotungstic acid for examination. Thin sections were also made using bacterial suspensions instead of bacteria adsorbed to grids. Flagella were easily observed with this indirect immunoperoxidase staining procedure. The anti-flagellin antibody was specific for flagella on all three serogroups of *L. pneumophila* that were investigated.

2. Immunoferritin-Labeled Antibody

Electron microscopy was used to identify Legionella in formalin-fixed and unfixed lung tissue.[5,28] Suspensions of bacteria from culture were also identified by this method. Ground tissue or culture suspensions were placed on formvar-carbon-coated grids and floated upside down on diluted rabbit antiserum prepared by immunizing a rabbit with *L. pneumophila*, serogroup 1. After incubation at 37°C for 30 min the grids were washed and floated on ferritin-conjugated anti-rabbit globulin diluted in buffer containing a wetting agent to reduce nonspecific reactions. After washing, the grids were dried and examined. Ferritin granules were observed on the surface of *Legionella* but not on other bacteria tested.

Serogroup-specific antigen was visualized on the surface of *L. pneumophila* using a similar technique.[29] Bacterial suspensions were fixed in glutaraldehyde, washed, and suspended in diluted rabbit anti-*Legionella* serum. After incubation at 23°C for 30 min, the suspensions were washed by centrifugation and incubated with ferritin-conjugated goat anti-rabbit IgG antibody for 30 min at 4°C. After washing, the suspensions were fixed, sectioned, and observed with an electron microscope. The ferritin granules were found on the surface of intact *Legionella*.

The immunoperoxidase and immunoferritin procedures are powerful research tools. Although antigens can be identified at specific locations on the bacteria, this may not be suitable for diagnostic purposes because the methods are time and cost intensive.

B. Light Microscopy

1. Bacterial Stains

Modified Gram stain — *Legionella* are often not stained by Gram's method.[30] This is particularly true with tissue specimens. However, peritoneal fluid samples from guinea pigs infected with *Legionella* were visible after Gram staining if neutral red was substituted for safranin.[31] This counter stain was left on the smear for 15 to 20 min instead of the usual 1 min. Although this procedure is not as sensitive as the direct fluorescent antibody method, it may be useful for screening clinical specimens from sites which do not have a resident normal flora.

Giemsa stain — A Giemsa stain has been used to demonstrate *Legionella* in smears of tissue scrapings, cultures, and embryonated egg-yolk-sac fluid.[32] However, the method may not permit visualization of the bacteria in tissue sections.

Silver stains — Silver impregnation stains have been successfully used to stain *Legionella* in tissue sections. The Dieterle stain[33] permitted visualization of *Legionella in situ*, but the Warthin-Starry silver stain appeared to work well and was much easier to perform.[30,34] This stain and the Geimsa stain have been used to screen tissue sections for bacteria such as *Legionella*.[35]

2. Modified Fluorescent Antibody Procedures

The direct fluorescent antibody (DFA) method is both sensitive and specific. However, background autofluorescence can make interpretation of results difficult. This may be encountered with lung tissue imprints or deparaffinized lung tissue sections from patients with extensive fibrosis. One method to reduce background fluorescence was to add crystal violet to the smear prior to performing the fluorescent antibody procedure.[36] Deparaffinized lung tissue sections on microscope slides were dipped quickly in diluted Gram's crystal violet, immediately washed in distilled water, air-dried, and stained by the FA procedure. If too much crystal violet remained on the smear, it was decolorized with alcohol before the FA stain was applied.

The presence of *L. pneumophila* in lung tissue was enhanced by an indirect fluorescent antibody (IFA) test.[37] Formalin-fixed lung tissue from one legionellosis patient was post-fixed in 1% osmium tetroxide for 1 hr, dehydrated with alcohol and propylene oxide, embedded, and polymerized overnight at 60°C in epoxy resin. Sections 1 μm thick were placed on slides, treated to remove the resin, washed in alcohol, and placed in water. After trypsinization for 5 min at 37°C and washing, rabbit anti-*Legionella* antibody was added to the slides and incubated for 30 min at room temperature. After washing, fluorescein-conjugated anti-rabbit globulin was added and the slides were incubated as before. After washing, the sections were mounted with glycerol and examined with a fluorescent microscope. Although this method may permit better visualization of *Legionella in situ* than the 5 min paraffin embedding technique, this method is available only to those laboratories that have experience with epoxy embedding and sectioning.

Flow cytometry reportedly can be used to detect *Legionella* in samples that contain at least 10^4 bacteria per milliliter.[38] Formalinized *L. pneumophila*, washed in buffered saline and adjusted to a density of about 10^8/mℓ were incubated for 4 hr at room temperature with fluorescein-conjugated rabbit anti-*Legionella* globulin. After centrifugation, pelleted *Legionella* were suspended in saline to a density of 10^7/mℓ and analyzed with a laser cell sorter (Coulter Electronics, Hialeah, Fla.) interfaced to a minicomputer. *Legionella* incubated with fluorescein-conjugated normal rabbit globulin served as controls. Specific fluorescent antibody-treated bacteria produced a characteristic broad peak of high-intensity fluorescence.

Escherichia coli and *Saccharomyces cerevisiae* did not produce this characteristic peak. Although this procedure may find application in certain situations, such as concentrated urine samples, greater sensitivity may be needed for diagnostic use with patient specimens.

3. Immunoassays

Glucose oxidase immunoassay — Another method to detect Legionella in deparaffinized tissue sections involved the use of a four-layer immunological sandwich technique.[39] Tissue sections were incubated for 30 min at room temperature with 3% normal goat serum. This was followed by incubation for 30 min with *Legionella*-specific rabbit antibody diluted in 1% normal goat serum. Sections were then incubated with dilute rabbit antiserum for 16 hr at 4°C. After washing, goat anti-rabbit IgG serum diluted in 1% normal goat serum was added to the smears and incubated for 15 min at room temperature. The smears were washed and incubated for 30 min in a solution containing soluble glucose oxidase complexed with rabbit anti-glucose oxidase antibody diluted in 1% goat serum. After washing, nitro blue tetrazolium was added. After incubation for 5 min at room temperature, the sections were washed in 4% buffered formaldehyde and stained with nuclear fast red plus phloxine. The method was reported to be specific for intact *Legionella* and also permitted visualization of antigen.[39] This four-layer sandwich technique did require several steps. Whether this method would be useful for rapid diagnosis has not been determined.

Peroxidase immunoassay — Lung sections, sputum samples, and cultures have been evaluated by a peroxidase-conjugated antibody technique. The method was briefly described, but no data were presented.[40] Formalin-killed *Legionella*, serogroups 1 to 4, were used to immunize mice. The IgG fraction of mouse ascites fluid was conjugated to horseradish peroxidase and used to detect *Legionella* in smears. After washing, bound peroxidase-conjugated IgG antibody was detected by adding the substrate, diaminobenzidine, followed by incubation, washing, and counterstaining with fast green. This method could detect *Legionella* which had been added to sputum samples. Mention was also made of an indirect immunoassay using rabbit anti-*Legionella* antibody followed by peroxidase-conjugated goat, anti-rabbit antibody. The sensitivity and specificity of these methods were not documented.

4. Tetrazolium Dye Reduction Test (INT)

The *Legionella* electron transport system reduced the dye, 2-*p*-iodophenyl-3-*p*-nitrophenyl-5-phenyl tetrazolium chloride (INT), to red formazan crystals which were observed with a light microscope.[41] This method, which detected metabolically active *Legionella*, was combined with the direct fluorescent antibody test to locate respiring *Legionella* in water samples. Tests were performed with cultures to determine the baseline reactivity of the INT test and its relationship to viable counts and culture optical density. INT was added to samples at a final concentration of 0.02 or 0.002% for cultures and water samples, respectively. After a 60-min incubation period, the reaction was stopped by adding formaldehyde at 0.1% final concentration. Samples were then placed on slides, stained by the FA technique, and observed with a combination of bright-field tungsten light and epi illumination UV light. This is a very interesting technique and may find wide application for testing water samples. However, its usefulness with patient specimens may be somewhat limited because the INT method may also detect election transport activity in phagocytic or tissue cells.

5. Sandwich Immunofluorescence Assay (SIA)

A novel method to detect *Legionella* soluble antigen was recently described.[42] The solid support used to detect antigen was prepared as follows. *Staphylococcus aureus*, Cowan 1, bacteria were suspended in 0.5% formalin PBS and held at room temperature for 3 hr. After washing, the suspension was incubated at 80°C for 5 min and cooled for 5 min in an ice bath. The staphylococci were then washed and suspended in buffered saline. In order to detect soluble *Legionella* antigen, protein A on the surface of the staphylococci was coupled to the Fc end of anti-*Legionella* IgG antibody. The staphylocci were placed on a microscope slide and heat fixed. Diluted antiserum produced by immunization of rabbits with heat-killed *Legionella* vaccine was placed on the smear and incubated at 37°C for 30 min. After the slide was washed and air-dried, soluble test antigen was placed on the slide and incubated as before. After washing, fluorescein-conjugated antibody to *Legionella* was added to the slide and incubated for 30 min. The slide was washed to remove unbound antibody and then observed with UV light microscope as in the standard FA procedure. If the test antigen corresponded to the serogroup specificity of the rabbit antiserum attached to the protein A on the staphylococci, the cocci fluoresced. This method detected soluble antigen with 100-fold more sensitivity than a similar agglutination test. Untreated urine collected from healthy volunteers mixed with soluble antigen from *Legionella* produced a positive reaction in the SIA test. This method may be quite useful for detection of soluble antigen in some patient samples.

C. Colorimetric and Radioimmunoassays

1. Enzyme-Linked Immunosorbent Assay (ELISA)

A direct ELISA was used to detect soluble *Legionella* antigen in patient specimens such as urine.[43] This colorimetric procedure was performed in polystyrene test tubes. Test tubes were coated with diluted rabbit anti-*Legionella* IgG by adding antiserum to the tubes and incubating for 1 hr at 37°C. The antibody was removed by aspiration, the tubes were rinsed, and 5% BSA was added. After incubation for 1 hr, the albumin was aspirated, the tubes were rinsed, and test urine sample was added to each tube and incubated as before. After aspirating the test urine and rinsing, dilute alkaline-phosphatase-conjugated anti-*Legionella* IgG antibody was added to each tube and incubated overnight at 4°C. After each tube was washed, *p*-nitrophenylphosphate and buffer were added. After incubation at 37°C for an appropriate period of time, the color reaction was stopped by adding sodium hydroxide. Color development was quantitated by spectrophotometry at a wavelength of 400 nm. Tests of urine specimens from 39 to 47 patients with legionellosis were positive. Urine from three patients with other bacterial diseases changed from positive to negative if their urine was heated to 100°C for 5 min before testing. Heating urine samples from legionellosis patients did not change the results of the ELISA tests. This was not the first application of the ELISA for detection of *Legionella* antigen; studies were performed previously with urine[44,45] and sputum[46] with mixed results. The assay system described here does seem to be specific and may have diagnostic applications.

2. Radioimmunoassay (RIA)

Test procedures similar to the ELISA described above used radionuclide-labeled antibody instead of enzyme-conjugated antibody as the indicator system. Detection of soluble *Legionella* antigen in patient urine by RIA used a procedure similar to that described above for ELISA.[47] After test urine was incubated for 1 hr at 37°C in polystyrene test tubes coated with rabbit, *Legionella*-specific IgG, the urine was aspirated and the tubes were rinsed. Iodinated ^{125}I-anti-*Legionella* IgG was added to the tubes and incubated for 1 hr at 37°C. The antibody was removed, the tubes were washed, and radioactivity was measured with a gamma-radiation spectrophotometer. Heat-treated RIA-positive urine samples remained positive. All urine specimens from patients with confirmed legionellosis were positive with this test if the patients had not been on antibiotic therapy for more than 2 days. However, use of higher-titered anti-*Legionella* rabbit IgG to coat the tubes may permit detection of urine antigen for periods greater than 2 days.[4] Only one false positive urine sample was reported. RIA tests of patient urine from 26 of 34 culture, DFA-, and/or IFA-confirmed legionellosis patients were reportedly positive.[48] The best RIA positive urine antigen results (93%) were obtained from patients that were culture positive. The RIA procedure may be a useful supplement to other tests for diagnostic laboratories that have a gamma counter.

A comparison of RIA and ELISA for detection of soluble antigen in urine indicated that the sensitivity of both procedures was similar.[43] However, the antigen incubation step for ELISA was overnight vs. 1 hr for RIA. Also, the ELISA did produce a few more false positive results than RIA unless test urine was heated. Depending on the equipment available to a diagnostic laboratory, ELISA or RIA may be used to screen for soluble antigen in patient urine samples. Of course, the ELISA avoids the procedures required for handling and disposing of radioactive nuclides.

D. Visible Agglutination or Precipitin Tests

1. Slide Agglutination Test (SAT)

Slide tests are rapid and do not require expertise to interpret, provided that the antibody used to identify the visible test antigen is specific. A slide test to identify the serogroup of *L. pneumophila* colonies from agar plates was recently published.[49] Serogroup-specific antiserum was produced by vaccinating rabbits with boiled *L. pneumophila*. Cultures to be tested were suspended in 10% neutral formalin. The agglutination test was performed by mixing one drop of test antigen with one drop of rabbit antiserum on a microscope slide and rocking for 1 min. Agglutination was scored from strong to barely visible. The test was serogroup-specific for *L. pneumophila* serogroups 1 to 4. The slide agglutination test has also been evaluated with four other species of *Legionella*.[42] The only reported departure from the previous protocol was to rock the mixture of undiluted rabbit antiserum and culture suspension for 15 sec instead of 1 min. A positive test was visible agglutination within 15 sec. Again, the tests were species and serogroup specific.

This test has also been used to study flagellated or nonflagellated live *Legionella*.[27] Two drops of bacteria were mixed with one drop of rabbit anti-flagellin from serogroup 1, *L. pneumophila*. Positive reactions showed visible agglutination within 15 min. The anti-flagellin serum agglutinated flagellated but not nonflagellated live *Legionella*. Heat-killed bacteria did not agglutinate with the serum.

2. Card Agglutination Test

This procedure was developed for titration of patient sera using known *Legionella* antigen.[19] It may be possible to reverse the protocol and test *Legionella* culture suspensions with diluted hyperimmune rabbit serum. Comparisons with the slide agglutination test could be performed to determine whether this is a usable procedure.

3. Slide Coagglutination Test (coSAT)

A coagglutination test was developed as a modification of the slide agglutination test to detect the presence of soluble *Legionella* antigen.[42] *Staphylococcus aureus*, Cowan 1 strain, with protein A on the surface was used as an amplification system so that reactions between soluble antigen and specific antibody produced visible agglutination of the staphylococci. Cultures of *Legionella* on CYE agar were suspended in 10% neutral formalin when tests were performed with particulate antigen. Soluble antigen was prepared by suspending cultures in tubes of distilled water and heating in a boiling water bath for 15 min. *Legionella*-specific antisera were obtained by immunization of rabbits with heat-killed vaccines. The staphylococci were cultured in broth medium, washed by centrifugation, suspended in 0.5% formalin in PBS, and incubated at room temperature for 3 hr. After washing, the staphylococci were suspended in buffered saline, heated for 5 min in an 80°C water bath, cooled for 5 min in an ice bath, washed, and suspended in saline. The coSAT test was performed by adding appropriately diluted anti-*Legionella* serum to the staphylococcal reagent and mixing one drop of this with one drop of particulate or soluble test antigen on a slide. Visible agglutination after 15 sec was considered a positive test. Tests with whole cells or soluble antigens were species and serogroup specific. The advantages of the coSAT over the direct slide agglutination test were use of less antiserum and the ability to detect *Legionella* soluble antigen. Normal urine seeded with soluble antigen produced a positive coagglutination reaction if the urine was first heated for 10 min in a boiling water bath. This procedure is less complicated than RIA or ELISA and can be performed without the need for sophisticated equipment.

4. Immunodiffusion Test

Legionella soluble antigens can be identified by an immunodiffusion test using wells cut in 1% agarose-coated slides.[50] Soluble antigen was isolated by suspending *Legionella* cultures in PBS and heating for 30 min in a boiling water bath. Soluble antigen was removed from the supernatant fluid after centrifugation of the boiled suspension. Anti-*Legionella* serum produced by vaccination of rabbits was placed in a central well and soluble antigens were placed in each of six surrounding wells. Precipitin bands were checked after 1, 2, and 3 days for identity, nonidentity, and partial identity. The method was also used to identify *Legionella* envelope antigen.[51] The test may be useful for studies of antigen identity between several isolates of *Legionella*. It is not known whether this test could be used to detect soluble antigen in patient specimens such as urine.

5. Counterimmunoelectrophoresis (CIE)

This test has been used to detect bacterial soluble antigen in cerebrospinal fluid. This method has been applied to detection of *Legionella* antibody in patient serum[20] as discussed previously, and identification of *Legionella* soluble antigens.[51] Loosely associated surface antigens from *L. pneumophila* serogroups 1 and 2 were isolated, partially purified, and analyzed by CIE.[51] Antiserum was produced by immunization of rabbits with crude surface antigen. Microscope slides were coated with 0.75% agarose, pH 8.6, two wells were punched across from each other, soluble test antigen was added to each well, and the slides were electrophoresed at 7 V/cm for 60 min at room temperature. Antiserum was placed in wells cut on each side of each antigen well and the slides were incubated at 37°C for 18 hr in a moist chamber. A major antigenic component migrated toward the cathode but a minor antigen migrated toward the anode. Later experiments were performed by placing antiserum in a well on the cathode side of the antigen well. For some unknown reason, the CIE test was less sensitive in this application than CIE diagnostic tests for other bacteria.

6. Two-Dimensional Crossed Immunoelectrophoresis

This test, along with immunoelectrophoresis, was used to study *Legionella* group-specific and common antigens.[52] Cultures grown in chemically defined medium were harvested by centrifugation, washed, and suspended in 0.25% formalin plus saline. After incubation overnight at 4°C, soluble test antigen was prepared by disruption in a Braun homogenizer in the presence of sodium deoxycholate. After concentration by pressure dialysis, soluble antigens were recovered in the supernatant fluids after centrifugation at 40,000 × g for 30 min. Before use, soluble antigens were mixed with Triton X-100 (5% final concentration). Anti-*Legionella* sera were produced by vaccination of rabbits. Two-dimensional crossed immunoelectrophoresis was performed on 4.3-cm square glass slides coated with 0.8% agarose. Antigen was placed in a well cut in the agarose and electrophoresed at 8 V/cm for 45 min. A layer of agarose containing diluted anti-*Legionella* serum was placed on the slide and electrophoresed at room temperature in the second direction for 16 hr at 1.5 V/cm. The slides were washed and stained with coomassie brilliant blue. *L. pneumophila* serogroups 1 to 4 produced 8 to 11 precipitin lines against homologous antisera. Both common antigens and serogroup-specific antigens were resolved by this method, but the latter were more clearly defined. There were no cross-reactions between serogroup-specific antigens. This procedure may be too rigorous for use as a diagnostic tool, but it does permit more detailed evaluation of *Legionella* antigens than heretofore possible.

7. Reversed Passive Hemagglutination Test (RPHA)

Detection of *Legionella* soluble antigen in urine from one patient with suspected legionellosis was accomplished with this test.[53] After the erythrocytes had been tanned with dilute tannic acid for 15 min at 37°C, they were incubated with diluted goat anti-*Legionella* globulin for 15 min at 37°C, washed, and suspended in buffered saline containing 1% normal rabbit serum. Test antigen samples were diluted eightfold in buffered saline containing normal rabbit serum, heated for 30 min at 60°C to destroy complement, and adsorbed with sheep red blood cells for 30 min to reduce nonspecific reactions. The test samples were serially diluted in microtiter wells and antibody-containing sensitized sheep red blood cells were added to each well. After incubation at room temperature for 2 to 3 hr, agglutination at a titer greater than 1:16 was considered positive for *Legionella* antigen. Multiple urine samples from one patient with confirmed legionellosis produced positive agglutination reactions. It is not known whether this procedure was more sensitive than some of the other tests described for detection of soluble antigen, but a detection limit of 4.3 ng protein per milliliter was reported.

8. Lectin Agglutination Test

Plant agglutinins were used to evaluate the agglutination patterns of *Legionella*.[54] Cultures from CYE agar were suspended in PBS, washed, and mixed with plant agglutinin in microtiter wells. After incubation at 4°C overnight, agglutination patterns were scored. The positive reactions were not serogroup-specific with *L. pneumophila* isolates or other species of *Legionella*. Two plant agglutinins, from *Aloe arborescens* and *Persia americana*, did react with most isolates of *L. pneumophila*. Several other lectins and plant agglutinins did not react. The authors suggested that *Legionella* may not have accessible carbohydrate groups on their surface because the two reactive agglutinins seem to bind protein rather than carbohydrate. At this time, plant agglutinin reactions do not have any diagnostic significance.

E. Conclusions

Tests for *Legionella* antigens range from simple slide agglutination tests to highly sophisticated techniques involving electron microscopy. The primary objective of a diagnostic laboratory is to have a rapid, specific, and sensitive assay for antigen in body fluids. At this time, there have not been enough comparisons of each procedure with the same patient specimens to provide an answer to which test works best. In terms of simplicity and quantity of reagents needed, the slide coagglutination test would appear to be quite useful. Equivocal reactions may be further evaluated by using the sandwich immunofluorescence assay. Once *Legionella* have been isolated on CYE agar, several of the visible agglutination or precipitin tests could be used for screening purposes.

IV. TESTS FOR *LEGIONELLA* ENZYMES

Most species of *Legionella* have been characterized by biochemical tests commonly available to some diagnostic microbiology laboratories.[55,56] Tests for β-lactamase, catalase, gelatin liquefaction, hippurate hydrolysis, nitrate reduction, and oxidase are discussed in Chapter 2. Although the above tests provide useful information, serological tests such as DFA are more specific. Therefore, tests for other biochemical activities have been investigated in an effort to find a rapid *Legionella*-specific procedure that would provide diagnostic laboratories with an alternative to serological tests.

A. Fluorogenic Substrate Cleavage

These tests rely on the hydrolysis of substrates that produce products that fluoresce under long-wavelength UV light. Five species of *Legionella* cleaved diacetylfluorescein, whereas *Staphylococci* and most Gram-negative bacteria, with the exception of *Proteus rettgeri*, did not.[55] This test was also positive with *L. jordanis*.[57] Briefly, suspensions of *Legionella* were incubated with substrate for 5 min at 20°C. A positive test suspension produced a yellow-green fluorescence under a long-wavelength UV lamp. Other fluorogenic substrates were tested, but they were positive with both *Legionella* and other Gram-negative bacteria.[55] Therefore, diacetylfluorescein may have potential application in the diagnostic laboratory.

B. Colorimetric and Other Enzyme Tests

The API ZYM system (Analytab Products, Plainview, N.Y.) containing several chromogenic substrates has been tested with several *Legionella* species.[58,59] *Legionella* agar cultures were suspended in saline and used to fill each capsule on the API ZYM strip. After incubation for 4 hr at 37°C in air supplemented with 5% CO_2, two color development reagents were added. After 5 min, color development was scored on a scale from 0 to 5. All six *Legionella* species tested produced color changes suggesting alkaline and acid phosphatase, leucine aminopeptidase, butyrate and caprylate esterase, and phosphoamidase. However, *L. bozemanii* was caprylate esterase negative. All carbohydrate substrates were negative.[59] Comparisons with other Gram-negative bacteria were not reported, but other genera do have some of these enzyme activities. Further studies will indicate whether this system can be used to identify *Legionella*.

A chromogenic assay with four para-nitroanilide (p-NA) conjugated peptides (KABI Diagnostica, Stockholm) has been used to analyze peptidase activity of five *Legionella* species.[60,61] Culture supernatants were dialyzed, concentrated by ultrafiltration, and placed in microtiter wells to assay enzyme activity.[60] Alternatively, a sonicated extract of *L. pneumophila* was used.[61] Buffer and conjugated substrates were added to each well and incubated at 37°C for up to 30 min. Release of yellow-colored para-nitroanilide due to peptidase activity was measured at 410 nm with a spectrophotometer. The assay was pH and crude enzyme protein dependent. Culture supernatants of *L. pneumophila* serogroups 1 to 6 were moderately active on the Bz-Ile-Glu-Gly-Arg-pNA substrate; the other species had lower levels of activity. Activity on three other substrates was considerably less.[60] Experiments with sonicated of *L. pneumophila* reportedly had no activity.[61] Concentrated culture supernatants from *L. pneumophila, L. bozemanii, L. dumoffi*, and *L. gormanii* were most active on SucOMe-Arg-Pro-Tyr-pNA substrate. *L. micdadei* was inactive on 18 different peptides.[61] The authors concluded that chymotrypsin-like activity predominated. At this time, there is not enough data to make conclusions about use of these assays for identification of *Legionella* species.

Other tests for proteolytic activity have been reported.[56,62,63] *L. pneumophila* serogroups 1 to 6 degraded denatured casein, skim milk, gelatin, and Remzol brilliant blue conjugated hide powder.[62] Bacteria were cultured in liquid medium and removed by centrifugation. Culture supernatants were filter sterilized and used directly for enzyme assays. Skim milk, gelatin, and casein assays were performed with culture plates. Hide powder protease reactions, the main subject of this report, were done in test tubes.[62] All substrate incubations were at 37°C for various time intervals. The hide powder proteolytic activity was temperature and pH dependent. The authors concluded that *L. pneumophila* produced a neutral protease with an approximate molecular weight of 40,000. Again, more studies will be needed to determine the relevance of these findings to laboratory identification of *Legionella*.

The specificity of *L. pneumophila* serogroup 1 caseinate precipitating enzyme was studied by using anti-proteinase antibody produced by rabbit vaccination.[64] Enzyme was isolated from the supernatant of broth cultures incubated at 37°C for 14 hr or at 22°C for 80 hr. Supernatant fluid was filter sterilized, dialyzed overnight, concentrated by ultrafiltration, and fractionated on a Sephadex G-100 chromatography column. Column fractions containing caseinate-precipitating activity were pooled, concentrated by ultrafiltration, and used to vaccinate rabbits. Assays were performed with agar plates containing 1% sodium casenate and 0.004% $MgCl_2$ at pH 6.2. Serial dilutions of caseinate-precipitating enzyme were added to paper strips and the strips were placed on the surface of the agar and incubated at 37°C for 4 to 18 hr. The highest dilution of enzyme producing visible precipitation was considered one unit. Specific antibody inhibition of caseinate precipitation was performed by placing strips containing antibody that had been subjected to paper electrophoresis on the surface of the agar and incubating for 3 hr at 37°C. The antibody strips were then removed from the agar and replaced by strips containing enzyme. *L. pneumophila* produced significant levels of caseinate-precipitating enzymes. Three other species of *Legionella* were much less active; *L. micdadei* had no activity. Antibody made against *L. pneumophila* caseinate-precipitating enzyme prevented enzyme activity of all *Legionella* species tested. Antibodies to proteinase enzymes isolated from several other Gram-positive and Gram-negative bacteria did not prevent *Legionella* caseinate-precipitating enzyme activity. Sera from patients recovering from legionellosis were tested for specific, anti-proteinase antibody to *Legionella* caseinate-precipitating enzyme but no inhibition was observed.[64]

Proteolytic activity on human serum proteins was determined by incubating a suspension of *L. pneumophila* with normal human serum for 4 hr at 37°C.[63] The serum was subjected to immunoelectrophoresis for 3 hr on slides containing 2% agar in dietylbarbiturate acetate buffer, pH 8.2. After electrophoresis, rabbit antihuman serum was added to determine if there were changes in the location of serum protein precipitin bands. Five serum proteins were affected; α_1-acid glycoprotein, α_1-antichymotrypsin, β-lipoprotein, β_{1E}-globulin, and β_2-glycoprotein-I.[63]

Toxin produced by *L. pneumophila*, *L. micdadei*, *L. bozemanii*, *L. dumoffii*, and *L. gormanii* reportedly have antigenic relatedness that can be identified by a precipitin test, according to a recent review.[65] Other tests of toxin activity included mouse macrophage viability assays, Chinese hamster ovary cell cytotoxicity tests and stimulation of polymorphonuclear leukocyte (PMNL) chemiluminescence. *Legionella* toxin may also impair human PMNL superoxide production.[66] These in vitro laboratory assays are usually not available in diagnostic laboratories, however. It is possible that further investigations of extracellular enzymes and/or toxins produced by *Legionella* species may result in development of a diagnostic test that could be used to screen patient sera for anti-enzyme antibody, or to identify *Legionella* colonies on agar plates.

V. CULTURE METHODS

Culture is routinely used for diagnosis of legionellosis. The basic agar culture method plus embryonated chicken egg yolk sac and guinea pig procedures were discussed in Chapter 3. This section will cover recent modifications of the standard agar culture technique. *Legionella* have been cultured from sputum, tracheal aspirate, bronchial wash, bronchial brushing, pleural fluid, lung, spleen and liver tissue, hemodialysis fluid and blood. Some specimens contain normal flora that inhibit the growth of *Legionella*.[67] Therefore, the basic BCYE agar developed by Feeley et al.[68] as modified by Pasculle et al.[69] has been supplemented with various antibiotics and/or dyes to enhance recovery of *Legionella*.

A. Antibiotic-Supplemented CYE

Antibiotics have been added to inhibit the growth of oropharyngeal flora while permitting growth of *L. pneumophila*. The first reported medium contained polymyxin B (40 units/mℓ) and vancomycin (0.5 μg/mℓ).[70] This medium was tested with three lung specimens contaminated with one or more of *Citrobacter*, *Klebsiella*, *Enterobacter*, *Staphylococcus*, and *Candida*. Contaminants in two of three specimens did not grow on the antibiotic-containing CYE medium. *Candida* was not inhibited but the addition of anisomycin (80 μg/mℓ) was reportedly effective.

A survey of various combinations of five antibiotics plus α-ketoglutarate and sodium selenate found that BCYE agar supplemented with α-ketoglutarate, anisomycin (80 μg/mℓ), polymyxin B (80 units/mℓ), and cefamandole (4 μg/mℓ) produced the best results.[71] Twelve patient specimens yielded significantly better recoveries of *L. pneumophila* on this antibiotic-containing medium, but *Pseudomonas* and group D streptococci were not inhibited.

B. Other Media

1. Diphasic Medium

Liquid specimens from body sites without a normal flora, such as blood or pleural fluid, may produce greater recovery rates of *Legionella* if diphasic culture medium is substituted for agar medium. Although such patient specimens have not been tested, this method deserves further investigation. Diphasic medium containing 50% (v/v) solidified CYE agar below an equal volume of liquid medium containing either corn starch instead of charcoal[72] or filter-sterilized yeast extract medium without charcoal or starch[73] have reportedly been useful.

2. Enriched Blood Agar

Medium containing 5% horse blood, blood agar base, L-cysteine, and ferric pyrophosphate, was reported to be useful for isolation of *L. pneumophila* from patient specimens.[74] However, more parallel comparisons with BCYE medium will be needed to confirm this observation.

3. Transparent Culture Medium

Charcoal in BCYE agar makes the medium opaque so that colonies cannot be easily observed with transmitted light under a culture plate. A clear agar medium (YPH) containing yeast extract, monosodium glutamate, potassium phosphate, soluble starch, L-cysteine, ferric pyrophosphate, and hemin reportedly produced better growth of *Legionella* than CYE agar.[75] The plating efficiency or recovery of numbers of colonies on CYE vs. YPH was similar. Again, parallel studies with primary patient specimens will indicate the usefulness of this medium for routine culture of *Legionella*.

4. Antibody Agar Medium

Colonies of *L. pneumophila* serogroups 1 to 6 and four other *Legionella* species developed precipitin rings when cultured on modified medium containing appropriate levels of anti-*Legionella* antibody.[76] Antiserum was produced by vaccination of rabbits with formalinized suspensions of *Legionella*. Liquid medium containing yeast extract, L-cysteine and ferric pyrophosphate was filter sterilized and mixed with 1% agarose (w/v final concentration) that had been sterilized by autoclaving. The medium was then mixed with anti-*Legionella* serum (5% v/v, final concentration) and allowed to solidify in a petri dish. Tissue samples from *Legionella* infected guinea pigs, water samples seeded with *L. pneumophila*, and culture suspensions all produced visible precipitin rings around individual *Legionella* colonies when the agar contained the appropriate antiserum. No precipitin rings were found around colonies of other Gram-negative bacteria. This procedure may facilitate diagnosis of legionellosis by culture if further studies indicate that similar results are obtained with patient specimens. However, anti-*Legionella* rabbit serum would be needed in larger quantities and antibiotic supplementation of the agar medium may still be required for isolation of *Legionella* from contaminated specimens.

C. Selective/Differential Culture Techniques

Isolation of *Legionella* from environmental samples has proven to be difficult because of the presence of high numbers of other bacteria which either inhibit the growth of *Legionella* or overgrow *Legionella* colonies. Methods that have been applied to isolation of environmental *Legionella* will be discussed because some methods may become applicable to patient samples contaminated with other bacteria.

1. Dye-Containing Agar

BCYE agar containing 0.001% bromcresol purple and 0.001% bromthymol blue permitted visual differentiation by color between colonies of *L. pneumophila* (pale green to white), *L. micdadei;* (green), and *L. bozemanii* (blue-gray).[77] Other genera of bacteria may produce colored colonies on this medium so additional methods should be used to verify *Legionella*.

BCYE agar supplemented with 0.01% aniline blue permitted visual differentiation between colonies of *L. pneumophila* and four other species of *Legionella*.[78] *L. pneumophila* colonies were colorless but produced a yellow-green fluorescence in the medium when exposed to long-wavelength UV light. Conversely, colonies of *L. micdadei, L. dumoffii, L. bozemanii*, and *L. gormanii* incorporated the blue dye. *L. dumoffii* colonies were dark blue, *L. bozemanii* colonies were blue, and *L. micdadei* or *L. gormanii* were light blue. When exposed to UV light, colonies of *L. bozemanii* and *dumoffii* produced yellow-green fluorescence in the medium and blue-white fluorescent colonies. Therefore, *Legionella* colonies on this medium can be presumptively identified by colony color, presence of UV fluorescence in the medium, and UV fluorescence of the colonies. Growth on the dye-containing medium was reported to be similar to that observed on BCYE agar without aniline dye.[78]

2. Antibiotic Enrichment Before Plating

Water samples seeded with *L. pneumophila* serogroups 1 to 4 were mixed with equal amounts of tryptic soy broth containing final concentrations of colistimethate (75 μg/mℓ), nystatin (125 units/mℓ), and vancomycin (30 μg/mℓ).[79] After incubation at room temperature for 8 hr, the samples were plated on agar medium. Numbers of contaminating bacteria were dramatically reduced but not eliminated by this procedure. Further studies are required to determine the suitability of this method for patient specimens.

3. Low pH Treatment Before Plating

In order to reduce the contaminating bacterial flora, water samples containing *Legionella* were concentrated by centrifugation, the sediments were suspended in 1 mℓ of water and added to 9 mℓ of buffer containing 0.2 *M* KCl and HCl, pH 2.2.[80] After incubation for 5 min at room temperature the samples were placed on CYE agar alone or on agar containing cephalothin (4 μg/mℓ), colistin (16 μg/mℓ), vancomycin (0.5 μg/mℓ), and cycloheximide (80 μg/mℓ). A survey of 11 environmental samples indicated that acid treatment and plating on antibiotic-containing medium usually enhanced recovery of *Legionella*.

Another investigation of *Legionella* in potable water suggested that acid treatment did not enhance recovery over direct plating on BCYE agar containing α-ketoglutarate (0.1%), cefamandole (4 μg/mℓ), polymyxin B (80 units/mℓ), and anisomycin (80 μg/mℓ),[81] or on BCYE agar supplemented with the above components plus glycine (0.3%), vancomycin (1 μg/mℓ), bromthymol blue (10 μg/mℓ), and bromcresol purple (10 μg/mℓ).[82] The latter was superior to the former medium for isolation of *Legionella* from potable water.[82] Clearly, no one method has proven to be superior for isolation of *Legionella* from water samples. The number and types of contaminating bacteria vary with the source of water. For example, the mean of bacteria in potable water is much lower than in cooling tower water from air-conditioning systems.[89]

4. Heat Treatment Before Plating

Heating respiratory tract specimens was tested to determine whether recovery of *Legionella* would be enhanced.[83] Seven DFA-positive but culture-negative patient specimens were heated in a water bath at 60°C for 1, 2, and 3 min, placed in ice water for 15 sec, and held at room temperature for plating. Heated respiratory tract specimens plated on antibiotic-containing BCYE medium were culture positive in 17 of 28 specimens vs. of 13 of 28 without heat treatment. However, heat treatments did not always significantly reduce the numbers of contaminating bacteria except for one specimen containing *P. aeruginosa*.[83] When all else fails, heat treatment should be considered.

5. Other Culture Modifications

L. pneumophila cultured from patient specimens and environmental sites were not inhibited by 0.3% glycine incorporated into BCYE agar medium.[84] However, this concentration of glycine did inhibit growth of other bacteria. Optimum recovery of *Legionella* was reported with BCYE agar containing 0.3% glycine, polymyxin B (100 units/mℓ), and vancomycin (5 μg/mℓ). However, a single isolate of *L. gormanii* did not grow on this medium. This medium has produced good isolation results from additional environmental samples.[85] Also, *Legionella* were isolated from environmental samples taken from potable water plumbing fixtures and plated on BCYE agar, or BCYE agar supplemented with polymxin B (50 units/mℓ), vancomycin (1 μg/mℓ), 0.001% bromthymol blue, and 0.001% bromcresol purple.[50]

D. Conclusions

There seems to be general agreement that BCYE agar medium is adequate for isolation of *Legionella* from patient specimens that do not contain other bacteria. However, the use of media supplemented with various antibiotics seem to produce variable isolation rates among the various laboratories where parallel tests have been performed. Further evaluation of antibiotics, medium without charcoal, antibody-containing medium, and dye-containing medium will be needed before definite conclusions can be reached on the applicability of these modifications to patient specimens.

Isolation of *Legionella* from environmental sources has been intensively investigated by several authors. Antibiotic or low pH treatments have increased recovery rates in some situations but not others. Antibiotic and/or amino acid supplementation of BCYE agar medium has been useful, especially for culture of potable water samples or associated plumbing fixtures. Culture media containing dyes have also permitted more rapid presumptive identification of species other then *L. pneumophila*. Although differential/selective culture media have been evaluated for over 5 years, most authors recommend that patient and environmental samples continue to be plated on BCYE agar medium in addition to one of the supplemented agar media.

VI. MAGNETOPHORETIC ENRICHMENT

Many of the new procedures for diagnosis of legionellosis have been adapted from methods that were successful with other bacteria. *Legionella* use amino acids instead of glucose as a source of carbon and energy.[86,87] *Legionella* are also antigenically and genetically different from other bacteria. However, these characteristics have not led to the development of a rapid laboratory test that will confirm the presence of low numbers of *Legionella* in patient specimens that are contaminated with other bacteria such as oropharyngeal flora. The DFA test, although specific, requires an average number of about 10^5 *Legionella* per milliliter of specimen before one fluorescent organism will be observed in each oil immersion field with a UV light microscope.

One approach to this situation has been development of magnetic particles,[88] smaller than bacteria, that contain anti-*Legionella* antibody on their surface. When such particles are placed in a sample containing *Legionella*, the particles attach to the surface of the bacteria, thus making them susceptible to a magnetic field. If the sample is placed near a strong magnetic gradient, *Legionella* will be pulled toward the magnet while other bacteria will not be affected. Nonmagnetic bacteria that do not have anti-*Legionella* magnetic particles bound to their surface can be removed, leaving behind a sample enriched for *Legionella*. The sample can then be plated on agar medium or tested by DFA or other rapid tests for specific antigen. In practice, this method has permitted an enrichment of approximately 10^4-fold with some samples.[90]

ACKNOWLEDGMENTS

The author wishes to thank Elizabeth Connors, Barbara Whitehouse, and especially Michelle O'Donnell for their editorial assistance with the manuscript.

REFERENCES

1. **Wilkinson, H. W. and Brake, B. J.,** Formalin-killed versus heat-killed *Legionella pneumophila* serogroup 1 antigen in the indirect immunofluorescence assay for legionellosis, *J. Clin. Microbiol.*, 16, 979, 1982.
2. **Wilkinson, H. W., Cruce, D. D., and Broome, C. V.,** Validation of *Legionella pneumophilia* indirect immunofluorescence assay with epidemic sera, *J. Clin. Microbiol.*, 13, 139, 1981.
3. **Wentworth, B. B. and Stiefel, H. E.,** Studies of the specificity of *Legionella* serology, *J. Clin. Microbiol.*, 15, 961, 1982.
4. **Zimmerman, S. E., French, M. L. V., Allen, S. D., Wilson, E., and Kohler, R. B.,** Immunoglobulin M antibody titers in the diagnosis of Legionnaires disease, *J. Clin. Microbiol.*, 16, 1007, 1982.
5. **Rodgers, F. G.,** Identification of *Legionella pneumophila* antigens and antibodies by immunoferritin electronmicroscopy, *J. Med. Microbiol.*, 15, 181, 1982.
6. **Farshy, C. E., Klein, G. C., and Feeley, J. C.,** Detection of antibodies to Legionnaires' disease organism by microagglutination and micro-enzyme-linked immunosorbent assay tests, *J. Clin. Microbiol.*, 7, 372, 1978.
7. **Vogel, F. R., Klein, T. W., Specter, S. C., Hitchings, M., and Friedman, H.,** Detection of antibodies to *Legionella pneumophila* in immune guinea pig serum by solid-phase immunofluorescence, *J. Clin. Microbiol.*, 13, 726, 1981.
8. **Thompson, T. A. and Wilkinson, H. W.,** Evaluation of a solid-phase immunofluorescence assay for detection of antibodies to *Legionella pneumophila*, *J. Clin. Microbiol.*, 16, 202, 1982.
9. **Black, C. M., Pine, L., Reimer, C. B., Benson, R. F., and Wells, T. W.,** Characterization of antibody responses in legionellosis with an immunofluorometric assay, *J. Clin. Microbiol.*, 15, 1077, 1982.
10. **Johnson, W., Elliott, J. A., Helms, C. M., and Renner, E. D.,** A high molecular weight antigen in Legionnaires' disease bacterium: isolation and partial characterization, *Ann. Intern. Med.*, 90, 638, 1979.
11. **Johnson, W., Pesanti, E., and Elliott, J.,** Serospecificity and opsonic activity of antisera to *Legionella pneumophila*, *Infect. Immun.*, 26, 698, 1979.
12. **Collins, M. T., Cho, S-N., and Reif, J. S.,** Prevalence of antibodies to *Legionella pneumophila* in animal populations, *J. Clin. Microbiol.*, 15, 130, 1982.
13. **Harrison, T. G. and Taylor, A. G.,** A rapid microagglutination test for the diagnosis of *Legionella pneumophila* (serogroup 1) infection, *J. Clin. Pathol.*, 35, 1028, 1982.
14. **Edson, D. C., Stiefel, H. E., Wentworth, B. B., and Wilson, D. L.,** Prevalence of antibodies to Legionnaires' disease, *Ann. Intern. Med.*, 90, 691, 1979.
15. **Yonke, C. A., Stiefel, H. E., Wilson, D. L., and Wentworth, B. B.,** Evaluation of an indirect hemagglutination test for *Legionella pneumophila* serogroups 1 to 4, *J. Clin. Microbiol.*, 13, 1040, 1981.
16. **Yonke, C. A., Stiefel, H. E., Wentworth, B. B., and Wilson, D. L.,** Prevalence of antibody to serogroups 1-4 of *Legionella pneumophila*, *Am. J. Epidemiol.*, 115, 633, 1982.
17. **Lennette, D. A., Lennette, E. T., Wentworth, B. B., French, M. L. V., and Lattimer, G. L.,** Serology of Legionnaires' disease; comparison of indirect fluorescent antibody, immune adherence, hemagglutination, and indirect hemagglutination tests, *J. Clin. Microbiol.*, 10, 876, 1979.
18. **Soriano, F., Aguilar, L., and Garces, J. L. G.,** Simple immunodiffusion test for detecting antibodies against *Legionella pneumophila* serotype 1, *J. Clin. Microbiol.*, 15, 330, 1982.
19. **Kleger, B. and Hartwig, R. A.,** Development and evaluation of a card agglutinin titer (CAT) test for serodiagnosis of Legionnaires' disease, *Publ. Health Lab.*, 38, 247, 1980.
20. **Holliday, M. G.,** The diagnosis of Legionnaires' disease by counterimmunoelectrophoresis, *J. Clin. Pathol.*, 33, 1174, 1980.
21. **Plouffe, J. F. and Baird, I. M.,** Lymphocyte transformation to *Legionella pneumophila*, *J. Clin. Lab. Immunol.*, 5, 149, 1981.

22. **Plouffe, J. F. and Baird, I. M.,** Lymphocyte blastogenic responses to *L. pneumophila* in acute legionellosis, *J. Clin. Lab. Immunol.*, 7, 43, 1982.
23. **Wong, K. H., Schalla, W. O., Arko, R. J., Bullard, J. C., and Feeley, J. C.,** Immunochemical, serologic, and immunologic properties of major antigens isolated from the Legionnaires' disease bacterium, *Ann. Intern. Med.*, 90, 634, 1979.
24. **Wong, K. H., McMaster, P. R. B., Feeley, J. C., Arko, R. J., Schalla, W. O., and Chandler, F. W.,** Detection of hypersensitivity to *Legionella pneumophila* in guinea pigs by skin test, *Curr. Microbiol.*, 4, 105, 1980.
25. **Elliott, J. A., Johnson, W., and Helms, C. M.,** Ultrastructural localization and protective activity of a high-molecular-weight antigen isolated from *Legionella pneumophila*, *Infect. Immun.*, 31, 822, 1981.
26. **Bohn, W.,** A fixation method for improved antibody penetration in electron microscopical immunoperoxidase studies, *J. Histochem. Cytochem.*, 26, 293, 1978.
27. **Elliott, J. A. and Johnson, W.,** Immunological and biochemical relationships among flagella isolated from *Legionella pneumophila* serogroups 1, 2, and 3, *Infect. Immun.*, 33, 602, 1981.
28. **Rodgers, F. G. and Macrae, A. D.,** Immunoferritin electronmicroscopy in legionellosis, *Lancet*, 1, 786, 1979.
29. **Flesher, A. R., Jennings, H. J., Lugowski, C., and Kasper, D. L.,** Isolation of a serogroup 1-specific antigen from *Legionella pneumophila*, *J. Infect. Dis.*, 145, 224, 1982.
30. **Dumoff, M.,** Direct in-vitro isolation of the Legionnaires' disease bacterium in two fatal cases, *Ann. Intern. Med.*, 90, 694, 1979.
31. **Tobin, J. O'H., Beare, J., Dunnill, M. S., French, M., Morris, P. J., Beare, J., Fisher-Hoch, S., Mitchell, R. G., and Muers, M. F.,** Legionnaires' disease in a transplant unit: isolation of the causative agent from shower baths, *Lancet*, 2, 118, 1980.
32. **Janssen, W. A. and Hedlund, K. W.,** Giemsa stain for *Legionella* pneumophila and TATLOCK bacterium, *Ann. Intern. Med.*, 94, 413, 1981.
33. **Van Orden, A. E. and Greer, P. W.,** Modification of the Dieterle spirochete stain, *J. Histotechnol.*, 1, 51, 1977.
34. **Pounder, D. J. and Stevens, S.,** Legionnaires' disease at autopsy, *Am. J. Forensic Med. Pathol.*, 2, 139, 1981.
35. **Pounder, D. J.,** Staining methods for legionella, *Hum. Pathol.*, 13, 185, 1982.
36. **Lowry, B. S., Vega, F. G., Jr., and Hedlund, K. W.,** Localization of *Legionella pneumophila* in tissue using FITC-conjugated specific antibody and a background stain, *Am. J. Clin. Pathol.*, 77, 601, 1982.
37. **Skinner, A. R. and Swann, A.,** The demonstration, by immunofluorescence, of *Legionella pneumophila* organisms in human lung tissue embedded in epoxy resin, *Histochemistry*, 71, 581, 1981.
38. **Ingram, M., Cleary, T. J., Price, B. J., Price, R. L., and Castro, A.,** Rapid detection of *Legionella pneumophila* by flow cytometry, *Cytometry*, 3, 134, 1982.
39. **Suffin, S. C., Kaufman, A. F., Whitaker, B., Muck, K. B., Prince, G. A., and Porter, D. D.,** *Legionella pneumophila*: identification in tissue sections by a new immunoenzymatic procedure, *Arch. Pathol. Lab. Med.*, 104, 283, 1980.
40. **Buschbaum, P. A., Cleary, T., Saldana, M., and Castro, A.,** Immunoperoxidase staining for the serotype-specific demonstration of *Legionella pneumophila*, *N. Engl. J. Med.*, 304, 613, 1981.
41. **Fliermans, C. B., Soracco, R. J., and Pope, D. H.,** Measure of *Legionella pneumophila* activity in situ, *Curr. Microbiol.*, 6, 89, 1981.
42. **Wilkinson, H. W. and Fikes, B. J.,** Detection of cell-associated or soluble antigens of *Legionella pneumophila* serogroups 1 to 6, *Legionella bozemanii*, *Legionella dumoffii*, *Legionella gormanii*, and *Legionella micdadei* by staphylococcal coagglutination tests, *J. Clin. Microbiol.*, 14, 322, 1981.
43. **Sathapatayavongs, B., Kohler, R. B., White, A., Winn, W. C., Jr., Girod, J. C., and Edelstein, P. H.,** Rapid diagnosis of Legionnaires' disease by urinary antigen detection: comparison of ELISA and radioimmunoassay, *Am. J. Med.*, 72, 576, 1982.
44. **Berdal, B. P., Farshy, C. E., and Feeley, J. C.,** Detection of *Legionella pneumophila* antigen in urine by enzyme-linked immunospecific assay, *J. Clin. Microbiol.*, 9, 575, 1979.
45. **Edelstein, P. H., Meyer, R. D., and Finegold, S. M.,** Laboratory diagnosis of Legionnaires' disease, *Am. Rev. Resp. Dis.*, 121, 317, 1980.
46. **Tilton, R. C.,** Legionnaires' disease antigen detected by enzyme-linked immunosorbent assay, *Ann. Intern. Med.*, 90, 697, 1979.
47. **Kohler, R. B., Zimmerman, S. E., Wilson, E., Allen, S. D., Edelstein, P. H., Wheat, L. J., and White, A.,** Rapid radioimmunoassay diagnosis of Legionnaires' disease, detection and partial characterization of urinary antigen, *Ann. Intern. Med.*, 94, 601, 1981.
48. **Kohler, R. B., Winn, W. C., Jr., Girod, J. C., and Wheat, L. J.,** Rapid diagnosis of pneumonia due to *Legionella pneumophila* serogroup 1, *J. Infect. Dis.*, 146, 444, 1982.
49. **Wilkinson, H. W. and Fikes, B. J.,** Slide agglutination test for serogrouping *Legionella pneumophila* and atypical *Legionella*-like organisms, *J. Clin. Microbiol.*, 11, 99, 1980.

50. **Garrity, G. M., Elder, E. M., Davis, B., Vickers, R. M., and Brown, A.,** Serological and genotypic diversity among serogroup 5-reacting environmental *Legionella* isolates, *J. Clin. Microbiol.*, 15, 646, 1982.
51. **Smith, R. A., DiGiorgio, S., Darner, J., and Wilhelm, A.,** Detection of *Legionella pneumophila* capsular-like envelope antigens by counterimmunoelectrophoresis, *J. Clin. Microbiol.*, 13, 637, 1981.
52. **Joly, J. R. and Kenny, G. E.,** Antigenic analysis of *Legionella pneumophila* and *Tatlockia micdadei (Legionella micdadei)* by two-dimensional (crossed) immunoelectrophoresis, *Infect. Immun.*, 35, 721, 1982.
53. **Mangiafico, J. A., Hedlund, K. W., and Knott, A. R.,** Rapid and sensitive method for quantitation of *Legionella pneumophila* serogroup 1 antigen from human urine, *J. Clin. Microbiol.*, 13, 843, 1981.
54. **Doyle, R. J., Nedjat-Haiem, F., Miller, R. D., and Keller, K. F.,** Interaction between plant agglutinins and *Legionella* species, *J. Clin. Microbiol.*, 15, 973, 1982.
55. **Orrison, L. H., Cherry, W. B., Fliermans, C. B., Dees, S. B., McDougal, L. K., and Dodd, D. J.,** Characteristics of environmental isolates of *Legionella pneumophila*, *Appl. Environ. Microbiol.*, 42, 109, 1981.
56. **Thorpe, T. C. and Miller, R. D.,** Extracellular enzymes of *Legionella pneumophila*, *Infect. Immun.*, 33, 632, 1981.
57. **Cherry, W. B., Gorman, G. W., Orrison, L. H., Moss, C. W., Steigerwalt, A. G., Wilkinson, H. W., Johnson, S. E., McKinney, R. M., and Brenner, D. J.,** *Legionella jordanis:* a new species of *Legionella* isolated from water and sewage, *J. Clin. Microbiol.*, 15, 290, 1982.
58. **Muller, H. E.,** Enzymatic profile of *Legionella pneumophila*, *J. Clin. Microbiol.*, 13, 423, 1981.
59. **Nolte, F. S., Hollick, G. E., and Robertson, R. G.,** Enzymatic activities of *Legionella pneumophila* and *Legionella*-like organisms, *J. Clin. Microbiol.*, 15, 175, 1982.
60. **Berdal, B. P., Hushovd, O., Olsvik, O., Odegard, O. R., and Bergan, T.,** Demonstration of extracellular proteolytic enzymes from *Legionella* species strains by using synthetic chromogenic peptide substrates, *Acta Pathol. Microbiol. Immunol. Scand. Sect. B*, 90, 119, 1982.
61. **Berdal, B. P., Olsvik, O., Myhre, S., and Omland, T.,** Demonstration of extracellular chymotrypsin-like activity from various *Legionella* species, *J. Clin. Microbiol.*, 16, 452, 1982.
62. **Thompson, M. R., Miller, R. D., and Iglewski, B. H.,** In vitro production of an extracellular protease by *Legionella pneumophila*, *Infect. Immun.*, 34, 299, 1981.
63. **Muller, H. G.,** Proteolytic action of *Legionella pneumophila* on human serum proteins, *Infect. Immun.*, 27, 51, 1980.
64. **Berdal, B. P. and Fossum, K.,** Occurrence and immunogenicity of proteinases from *Legionella* species, *Eur. J. Clin. Microbiol.*, 1, 7, 1982.
65. **Hedlund, K. W.,** *Legionella* toxin, *Pharmacol. Ther.*, 15, 123, 1981.
66. **Friedman, R. L., Lochner, J. E., Bigley, R. H., and Iglewski, B. H.,** The effects of *Legionella pneumophila* toxin on oxidative processes and bacterial killing of human polymorphonuclear leukocytes, *J. Infect. Dis.*, 146, 328, 1982.
67. **Flesher, A. R., Kasper, D. L., Modern, P. A., and Mason, E. O., Jr.,** *Legionella pneumophilia:* growth inhibition by human pharyngeal flora, *J. Infect. Dis.*, 142, 313, 1980.
68. **Feeley, J. C., Gibson, R. J., Gorman, G. W., Langford, N. C., Rasheed, J. K., Mackel, D. C., and Baine, W. B.,** Charcoal-yeast extract agar: primary isolation medium for *Legionella pneumophila*, *J. Clin. Microbiol.*, 10, 437, 1979.
69. **Pasculle, A. W., Feeley, J. C., Gibson, R. J., Cordes, L. G., Myerowitz, R. L., Patton, C. M., Gorman, G. W., Carmack, C. L., Ezzell, J. W., and Dowling, J. N.,** Pittsburgh pneumonia agent: direct isolation from human lung tissue, *J. Infect. Dis.*, 141, 727, 1980.
70. **Edelstein, P. H. and Finegold, S. M.,** Use of a semiselective medium to culture *Legionella pneumophila* from contaminated lung specimens, *J. Clin. Microbiol.*, 10, 141, 1979.
71. **Edelstein, P. H.,** Improved semiselective medium for isolation of *Legionella pneumophila* from contaminated clinical and environmental specimens, *J. Clin. Microbiol.*, 14, 298, 1981.
72. **Weiss, E., Peacock, M. G., and Williams, J. C.,** Glucose and glutamate metabolism of *Legionella pneumophila*, *Curr. Microbiol.*, 4, 1, 1980.
73. **Ristroph, J. D., Hedlund, K. W., and Allen, R. G.,** Liquid medium for growth of *Legionella pneumophila*, *J. Clin. Microbiol.*, 11, 19, 1980.
74. **Greaves, P. W.,** New methods for the isolation of *Legionella pneumophila*, *J. Clin. Pathol.*, 33, 581, 1980.
75. **Johnson, S. R., Schalla, W. O., Wong, K. H., and Perkins, G. H.,** Simple transparent medium for study of Legionellae, *J. Clin. Microbiol.*, 15, 342, 1982.
76. **Janssen, W. A. and Hedlund, K. W.,** Antiserum-agar plate method for simultaneous detection and direct isolation of *Legionella* species in clinical and environmental specimens, *J. Clin. Microbiol.*, 15, 1176, 1982.
77. **Vickers, R. M., Brown, A., and Garrity, G. M.,** Dye-containing buffered charcoal-yeast extract medium for differentiation of members of the family Legionellaceae, *J. Clin. Microbiol.*, 13, 380, 1981.

78. **Holmes, R. L.,** Aniline blue-containing buffered charcoal-yeast extract medium for presumptive identification of *Legionella* species, *J. Clin. Microbiol.*, 15, 723, 1982.
79. **Thorpe, T. C. and Miller, R. D.,** Negative enrichment procedure for isolation of *Legionella pneumophila* from seeded cooling tower water, *Appl. Environ. Microbiol.*, 40, 849, 1980.
80. **Bopp, C. A., Sumner, J. W., Morris, G. K., and Wells, J. G.,** Isolation of *Legionella* spp. from environmental water samples by low-pH treatment and use of a selective medium, *J. Clin. Microbiol.*, 13, 714, 1981.
81. **Edelstein, P. H., Snitzer, J. B., and Finegold, S. M.,** Isolation of *Legionella pneumophila* from hospital potable water specimens: comparison of direct plating with guinea pig inoculation, *J. Clin. Microbiol.*, 15, 1092, 1982.
82. **Edelstein, P. H.,** Comparative study of selective media for isolation of *Legionella pneumophila* from potable water, *J. Clin. Microbiol.*, 16, 697, 1982.
83. **Edelstein, P. H., Snitzer, J. B., and Bridge, J. A.,** Enhancement of recovery of *Legionella pneumophila* from contaminated respiratory tract specimens by heat, *J. Clin. Microbiol.*, 16, 1061, 1982.
84. **Wadowsky, R. M. and Yee, R. B.,** Glycine-containing selective medium for isolation of Legionellacae from environmental specimens, *Appl. Environ. Microbiol.*, 42, 768, 1981.
85. **Yee, R. B. and Wadowsky, R. M.,** Multiplication of *Legionella pneumophila* in unsterilized tap water, *Appl. Environ. Microbiol.*, 43, 1330, 1982.
86. **Tesh, M. J. and Miller, R. D.,** Amino acid requirements for *Legionella pneumophila* growth, *J. Clin. Microbiol.*, 13, 865, 1981.
87. **George, J. R., Pine, L., Reeves, M. W., and Harrell, W. K.,** Amino acid requirements of *Legionella pneumophila*, *J. Clin. Microbiol.*, 11, 286, 1980.
88. **Kronick, P. L., Campbell, G. L., and Joseph, K.,** Magnetic microspheres prepared by redox polymerization used in a cell separation based on gangliosides, *Science*, 200, 1074, 1978.
89. **Gilpin, R.W., Dillon, S. B., Keyser, P., and Androkites, A., Berube, M., Carpendale, N., Skorina, J., and Hunley, J.,** Disinfection of circulating water systems by ultraviolet light and halogenation, *Water Res.*, in press.
90. **Kronick, P. and Gilpin, R. W.,** Use of superparamagnetic particles for isolation of cells, *J. Biochem. Biophys. Meth.*, in press.

Part II: Epidemiology of Legionellosis

Chapter 5

LEGIONELLOSIS: EPIDEMIOLOGY

Ellen J. Mangione and Claire V. Broome

TABLE OF CONTENTS

I. INTRODUCTION

Since 1976, when Legionnaires' disease and its causative agent *Legionella pneumophila* were first recognized,[1,2] knowledge about the epidemiology of legionellosis has increased dramatically as a result of numerous investigations. Legionnaires' disease is now recognized as only one of two distinct clinical illnesses caused by the bacteria in the family Legionellaceae, genus *Legionella*, which now includes nine named species. The term "legionellosis" has been suggested to describe all infections due to this genus.

It is useful to divide cases of legionellosis into those which occur in an epidemic setting and those which occur sporadically. While epidemic investigations offer the opportunity to implicate a specific source of exposure to the organism in a defined population, sporadic case investigation allows for definition of characteristic exposures among large groups of independently occurring cases. It is important to remember, however, that defining cases as either epidemic or sporadic is somewhat artificial and, in fact, the occurrence of disease may represent a spectrum from sporadic cases to major epidemics.

A total of approximately 30 epidemics or outbreaks of legionellosis have been recognized. Outbreaks fall into 1 of the following 4 patterns:

1. Epidemics of pneumonic legionellosis
2. Epidemic nonpneumonic illness (Pontiac fever)
3. Hyperendemic pneumonic illness
4. Epidemic illness superimposed on hyperendemic disease

Characteristics of some of the outbreaks investigated by the Centers for Disease Control (CDC) are summarized and categorized in Table 1.

Although legionellosis and its etiologic agents have been recognized only recently, *Legionella* species are not new. *L. micdadei* which includes strains previously refered to as Tatlock,[20] HEBA, [21] and the Pittsburgh pneumonia agent, [22,23] and *L. pneumophila* (OLDA)[24]were first isolated in the 1940s, and *L. bozemanii* (WIGA)[21]in the 1950s, the latter two from sporadic cases. Five outbreaks of legionellosis are known to have occurred between 1947 and 1976—in 1957,[3] 1965,[4] 1968,[11] 1973,[12] and 1974[5]— and were investigated although a causative agent was not isolated. Retrospectively, these outbreaks, which are described below, were attributed to *Legionella*. Thus, legionellosis first appeared well before the 1976 Philadelphia outbreak in the form of both sporadic and epidemic illness.

The purpose of this chapter is to discuss the epidemiology of epidemic and sporadic illness caused by *Legionella*, including the reservoir and methods of transmission of the organism, risk factors for development of disease, and finally, the incidence and prevalence of disease.

II. EPIDEMIC LEGIONELLOSIS

Although first described in the U. S., legionellosis is in fact a cosmopolitan illness, with outbreaks described in Great Britain, Spain, The Netherlands, Germany, Sweden, France, Canada, and Italy.

Epidemic legionellosis exhibits seasonality, with about 20 of the major explosive outbreaks having occurred in July, August, or September, although cases associated with continuing nosocomial outbreaks have occurred throughout the year.

The most important and frequently recognized form of legionellosis is Legionnaires' disease. Legionnaires' disease is a severe pneumonic illness associated with high fever, diarrhea, and nonproductive cough. The incubation period for this illness is 2 to 10 days, with an attack rate of 0.2 to 5.0% and a case-fatality ratio of approximately 15%. Higher case-fatality ratios have been described during nosocomial outbreaks. Legionnaires' disease

Table 1
EPIDEMIOLOGIC ASPECTS OF LEGIONELLOSIS OUTBREAKS IN THE UNITED STATES, 1957—1981

	Dates	Number of cases	Attack rate (%)	Case fatality rate (%)
Epidemic pneumonic legionellosis				
Austin, Minn.[3]	June-August 1957	78	3.2	3
Washington, D. C.[4]	July-August 1965	81	1.4	17
Philadelphia, Pa.[5]	September 1974	20	2.9	10
Philadelphia, Pa.[1]	July-August 1976	221	6.8	15
Atlanta, Ga.[6]	July 1978	8	1-2	0
Dallas, Texas[7]	August-September 1978	18	1.2	11
Eau Clair, Wiss.[8]	June-July 1979	13	1.0	31
Memphis, Tenn.[9]	August-September 1978	39	2.8	13
Jamestown, N.Y.[a]	July 1979	4	4.0	25
Chicago, Ill.[10]	December 1979-March 1980	5	—	40
Pontiac fever				
Pontiac, Mich.[11]	July 1968	144	95	0
James River, Va.[12]	August 1973	10	100	0
Epidemic illness superimposed on hyperendemic disease				
Bloomington, Ind.[13]	January 1977-present	(39)[b]	0.2	(10)
Kingsport, Tenn.[14]	August-September 1977	(33)	0.6	(9)
New York, N.Y.[15]	August-September 1979	38	—	(8)
Burlington, Vt.[16]	May-October 1977	(69)	—	(25)
Los Angeles, Ca.[17]	May 1977-present	(49)	0.5	(31)
Hyperendemic pneumonic illness				
Norwalk, Conn.[18]	December 1976-present	28	0.7	38
Columbus, Ohio[19]	July 1977-present	(15)	—	(7)

[a] Shands, K., personal communication, 1979.
[b] ( ) Ongoing.

exhibits a male:female ratio of 2:1 and age specific attack rates that increase progressively with age over 50.

Several factors in addition to age and sex have been shown to increase risk of developing disease. The risk of Legionnaires' disease is increased in those who smoke. In the 1976 Philadelphia outbreak, the relative risk of disease was 3.4 for delegates who smoked cigarettes at the time of the convention.[1] It was also a significant risk factor in the 1977 outbreak in Burlington, Vermont[16] and among sporadically occurring cases.[25] Although, no statistically significant differences were found between case delegates and controls in the number of alcoholic drinks consumed during the convention in the Philadelphia outbreak, heavy alcohol consumption (3 or more drinks per day) was found to be a risk factor among sporadic cases[25] and among cases occurring in a nosocomial outbreak.[18] Other factors that have been shown to contribute to risk of disease were underlying illnesses, such as diabetes mellitus, congestive heart failure, malignancy and obstructive pulmonary disease, and use of immunosuppressive medication.

Pontiac fever, a second form of legionellosis, is a nonpneumonic flu-like illness that resolves without therapy. It may be difficult to recognize on the basis of clinical characteristics because of its similarity to viral syndromes; however, epidemics have been readily recognized based on characteristic epidemiologic features, including a short incubation period with a median of 36 hr, an attack rate of 90 to 100%, and a case-fatality ratio of 0%.

The large discrepancy between attack rates in outbreaks of Legionnaires' disease compared with those of Pontiac fever suggests the relative importance of host risk factors for each. In Pontiac fever, the entire population develops illness and host factors are of less significance, whereas only a small percentage of the population develops pneumonic legionellosis, even in a major outbreak.

III.RESERVOIR AND METHODS OF TRANSMISSION

Since the Philadelphia outbreak in 1976, investigators have sought the environmental sources of *L. pneumophila*. It has become apparent that legionellae are widespread in the environment, having been isolated from water from lakes, streams and riparian soil,[26] plumbing systems,[28] cooling towers,[9] and evaporative condensers.[27] The ubiquity of the organism implies that isolating *Legionella* from the environment in an outbreak does not establish a causal role. For example, during an investigation of cases of Legionnaires' disease at the Wadsworth Medical Center, water from an adjacent cooling tower was found to be culture positive for *L. pneumophila*.[17] Despite the institution of continuous chlorination of cooling tower water and repeated negative cultures, cases continued to occur.

The 1976 Philadelphia outbreak of Legionnaires' disease resulted in an exhaustive investigation into the source and transmission of illness. The major modes of spread that were investigated included foodborne, person-to-person, airborne, and waterborne.[1]

Since neither eating at any specific restaurant or bar, nor attendance at convention events at which food was offered was significantly associated with illness, foodborne transmission was not likely to be important in the spread of disease. Evidence gathered in subsequent investigations has supported this finding.

The possibility of person-to-person transmission was also evaluated in Philadelphia. Roommates and family members of both cases and controls were questioned about illness. Roommates of cases were not significantly different from roommates of controls in the likelihood of developing illness (AR* = 8.5%, 5 of 59 and AR = 7.2%, 5 of 69, respectively). Futhermore, 24 days after the convention, none of the contacts of ill delegates (140 adults and 53 children) of contacts of well delegates (143 adults and 68 children), who were interviewed, but contracted illness that met the clinical criteria of a case. Therefore, there was no evidence of person-to-person transmission.

Delegates to the 1976 Philadelphia convention were questioned about ingestion of water and ice at the Bellevue-Stratford Hotel. Although more Legionnaires who became ill drank water compared with delegates who remained well ($p = 0.1$), the significance of this observation disappeared when this variable was adjusted for time spent in the hotel lobby. An important risk factor for the development of LD in the Philadelphia outbreak in 1976 was exposure to the Bellevue-Stratford Hotel $p < 0.01$) Though no clustering within bedrooms could be demonstrated, attendance in a 14th floor hospitality room was found to be a risk factor. In addition, case delegates, including both delegates residing at the Bellevue-Stratford and those who stayed elsewhere, also were found to have spent more time in the hotel lobby than did those who remained well. Of note was the occurrence of illness among those who never entered the hotel but simply stood outside. Though not conclusive, these data are highly suggestive of airborne transmission of *Legionella*.

Investigations of other outbreaks have supported the hypothesis that airborne transmission

* Attack rate.

of *Legionella* occurs. In particular, cooling towers[9] evaporative condensers,[6,11,28] and an excavation site [4] have been implicated as the source of the organism.

A cooling tower is a unit designed to dissipate heat from water into the environment. Warm water is first pumped to the top of the cooling tower, then allowed to flow down over slats of wood or other material producing a spray. As the water falls, evaporation occurs, with a consequent drop in temperature. To facilitate the process, large fans blow air over the water droplets. As a result, exhaust air containing both the evaporated water and water droplets is disseminated from the tower.

Evidence implicating a cooling tower in the transmission of legionellosis was obtained during investigation of an outbreak of Legionnaires' disease in Memphis.[9] In early August 1978, heavy rains occurred in Memphis and caused flooding of the ground floor of one of the city's hospitals. Pumps supplying the main cooling tower were inactivated, and an auxiliary cooling tower that had not been subjected to routine treatment with biocides was put into service. During the subsequent 7 weeks, 44 cases of Legionnaires' disease occurred in Memphis—a city without previous confirmed cases of Legionnaires' disease. Thirty-nine of the 44 cases occurred in persons who had had contact with the hospital, either as patients, employees, visitors or passers-by, during the 10 days before onset of illness. All had been in or in front of the hospital while the auxiliary cooling tower was in use.

In addition to the temporal association with use of the auxiliary cooling tower, illness was significantly associated with hospitalization in areas of the hospital supplied by air intakes near that cooling tower. Tracer smoke studies demonstrated that contaminated aerosols from the tower could easily reach these intakes and the street below. *Legionella* was also isolated from the cooling tower.

In June and July 1979, an outbreak of LD involving 13 cases occurred in a hotel in Eau Claire, Wisconsin, in persons exposed to a meeting room partially ventilated by a fireplace with an open flue.[8] Smoke tracer studies demonstrated that the exhaust from a cooling tower located 5 m upwind flowed down the chimney and into the implicated meeting room, a room shown to be under negative pressure. The organism was isolated from both the cooling tower water and water on top of the implicated hotel building.

Evaporative condensers have also been implicated in the transmission of legionellosis in outbreaks that occurred in Pontiac, Michigan, and Atlanta, Georgia.[6] Evaporative condensers operate on a principle similar to a cooling tower. However, instead of the water passing over slats, it passes over a coil housing refrigerant. The refrigerant begins to flow through the coil as a hot gas, which is cooled, condensed, and recirculated, as water and air pass over the coils.

The earliest outbreak of Pontiac fever occurred at the Pontiac facility of the Oakland County Health Department in 1968, and provided convincing evidence of airborne transmission of the organisms from an evaporative condenser.[11] The first cases occurred on July 1. By July 3, 89 of the facility's 100 employees and 15 visitors had become ill. Three investigators worked in the building over the weekend without using the air conditioning unit. On July 8, the building was reopened and the air conditioning turned back on. The investigators became ill Tuesday night.

Cases had no common exposures outside of the health department building. A survey in the surrounding buildings of the county service complex and in the community revealed no increase in incidence of illness. In addition, many visitors to the building developed illness without having ingested food or water supplied in the building. No particular area of the building posed more of a threat of illness than other areas. Attack rates were higher in persons exposed during the morning than persons exposed for the first time during the afternoon, suggesting an increased risk of illness immediately after the air conditioning system was turned on.

An investigation of the air conditioning system revealed that one air conditioning system cooled the entire building. The air circulated to the building by this unit was readily contaminated with air and water from the adjacent evaporative condenser, either by leakage from the condenser duct to the adjacent ventilation duct or by airflow from the condenser exhaust vent to the intake vents for the air conditioner located in proximity on the roof. Although the etiology of the outbreak was not determined in 1968, guinea pigs exposed to the health department building and to laboratory aerosols of evaporative condenser water developed bronchopneumonia.[28] In 1978, *L. pneumophila* was isolated from stored lung tissue taken from these animals. In 1976, 37 paired sera obtained in 1968 from cases in the Pontiac outbreak and 10 controls were tested for antibody to *L. pneumophila* by indirect fluorescent antibody (IFA). A fourfold rise in titer to *L. pneumophila* was observed in 86.5% (32 of 37) of the cases, but none of the controls.

A second outbreak associated with an evaporative condenser occurred in July 1978 at a country club in Atlanta, Georgia.[6] Eight cases with clinical or radiological evidence of pneumonia developed during a 6-day period among club members. All cases were in persons who had played golf 2 to 10 days before becoming ill. In cases, the persons had played significantly more golf during that period than controls chosen from other members of the club. Two factors may have resulted in the outbreak. First, a rise in mean daily temperature from 21 to 30°C during the last 2 weeks of June would have resulted in an increased work load for the evaporative condenser and increased aerosol production. Second, during the last week of June and first week of July, winds were primarily from the west northwest, which would have directed the aerosol toward the 10th and 16th greens.

It is of interest that few outbreaks of legionellosis have occurred in winter and spring, when cooling towers and evaporative condensers are at their lowest usage level.

In 1973, a legionellosis outbreak associated with aerosolization of contaminated water occurred among men who cleaned a steam turbine condenser with compressed air.[12] The condenser was ordinarily cooled with water from the James River. All 10 of the men became ill with the typical picture of Pontiac fever within 2 days of exposure.

An outbreak without an apparent association with a ventilation system occurred in 1965, at a 6,000-bed psychiatric hospital in Washington, D.C.[4] The buildings were without air conditioning and there were no cooling towers on the grounds . A total of 81 patients met the case definition, including 55 with radiologic documentation of pneumonia (AR = 1.4%). Factors which increased the risk of illness included having ground privileges ($p < 0.0001$), and sleeping next to a window. Excavation work on the grounds had begun in early July. In mid July, heavy rains occurred, accompanied by winds of up to 47 mph—a velocity sufficient to uproot several trees on the property. The attack rate among residents living in buildings within 20 feet of an excavation site was significantly higher than for residents of buildings not immediately adjacent to an excavation site. Furthermore, 2 peaks of the epidemic curve followed earth moving at excavation sites close to the buildings on July 30 and August 4.

The Pontiac Health Department outbreak may also have been related to excavation. Five weeks before the Pontiac outbreak, ground adjacent to the building was regraded and paved and a large amount of dust was raised in the process. One week before the outbreak, torrential rains were followed by more than a 10°C increase in temperature to a mean of 27°C. It was therefore likely that the unit was turned on for the first time that season immediately before the outbreak, presumably shortly after the system was contaminated with the organism present in the soil.[53]

The role of contaminated potable water as a risk factor in the development of LD has been suggested by several investigations. The Wadsworth VA Hospital, completed in 1977, was the site of a continuing outbreak of LD for more than 3 years.[29] Though exposure to potable water could not be linked statistically to development of disease, *L. pneumophila*, serogroups 1 and 4 were isolated from the water, the same serogroups isolated from persons with legionellosis at the hospital. In March 1980, the number of LD cases increased dramatically from a baseline of about 4 cases per month, to 24 cases. Just before the March outbreak, the hospital's emergency pump had been tested. The emergency pump failed, causing a precipitous fall in water pressure throughout the hospital's potable water system. Though the problem was corrected in minutes, after the pressure drop the water remained distinctly discolored for weeks. Subsequently, a study was conducted to determine the effect of a similar pressure drop on the concentration of *Legionella* in the water in an area of the hospital not occupied by patients. Though *Legionella* had been isolated from the water before the study, cultures showed a 30-fold increase in the concentration of organisms in the water, and an increase in the number of taps culture-positive from 30% (3 of 10) to 50% (14 of 28).

Similarly, a cluster of nosocomial cases of pneumonic legionellosis due to *L. pneumophila* serogroup 1 occurred in a hospital in Paris, France. An association was demonstrated between cases and potable water known to be contaminated with the same subtype of serogroup 1 as had been found in a case isolate ($p = 0.003$).[54]

Although this evidence suggests that transmission can result from contaminated potable water, the organism may be present in the absence of associated disease. Tobin et al. conducted a survey of water systems in hospitals and hotels.[27] Forty-two isolates of *L. pneumophila* were made from 31 establishments. Six establishments had been associated with cases of LD. The other 25, however, had no known association with disease. Furthermore, in an investigation of nosocomial Legionnaires' disease in Chicago, *L. pneumophila* was found in showerheads.[10] The difficulty of interpreting this result is underscored by similar isolations of the organism from showerheads in control hospitals where no associated illness was detected.

IV. SPORADIC CASES

With the increased availability of laboratory methods capable of confirming the diagnosis, it has become evident that sporadic cases of Legionnaires' disease are more common than epidemic cases. Approximately twice as many sporadic cases as epidemic cases have been reported. The ratio of sporadic-to-epidemic cases is, in fact, probably higher given the problem of underreporting of sporadic cases to a passive surveillance system. It has been estimated that only between 0.8 and 2% of cases are reported to the surveillance system.

The Legionnaires' disease surveillance system was established in 1978. Since that time, 2113 cases of sporadic illness have been reported to the Centers for Disease Control (CDC). The total excludes cases known to have been associated with an epidemic. Though sporadic cases occur throughout the year, there is a seasonal variation similar to that of epidemics. Most occurred in the summer and fall with a peak during September. Figure 1 shows the average incidence of reported Legionnaires' disease per year for the period 1978 to 1981. Cases have been reported from every state, with a reported incidence ranging from less than 0.25 to more than 1.0 cases per 100,000 population (Figure 2). Sixty-six percent of cases were in persons ≥50 years old.

In 1979, Storch et al.[30] conducted a case-control study of 100 sporadic, community-acquired cases of LD reported to the CDC surveillance system. The cases were compared with three control groups: a clinical control group chosen from among patients with negative serodiagnostic tests for LD, and an acquaintance control group for both cases and clinical controls. In cases, the persons smoked more cigarettes, consumed more alcohol, and were more likely to reside near an excavation site than were controls.

A second study described the characteristics of the first 1,000 cases of sporadic LD reported to the surveillance system;[25] 19% were fatal. Seventy-five percent occurred in the period June to October. Ninety-eight percent were associated with pneumonia documented by chest

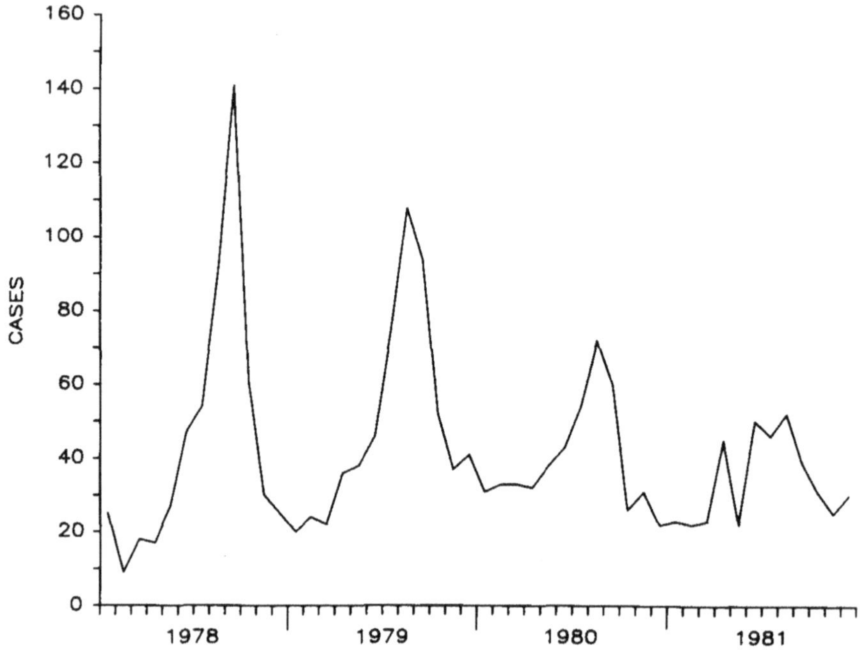

FIGURE 1. Legionellosis reported sporadic cases by month of onset, U. S. 1978 to 1981.

radiograph. Additional risk factors for sporadic disease included age over 50, having renal disease requiring dialysis or transplantation (RR* = 137), chronic obstructive pulmonary disease (RR = 20), diabetes mellitus (RR = 19), or cancer (RR = 94). Persons who smoke (RR = 2.3) and those being treated with immunosuppressive drugs (RR = 2.7) were also at increased risk of illness. Males had a relative risk of 2.6 compared to females. Of the 507 patients for whom information was available, 481 (95%) were probably community acquired since they were hospitalized no more than 2 days before onset of illness. The remaining 26 must be considered likely to have been nosocomial since all had been hospitalized 2 to 10 days before onset of illness.

Biases may be introduced into a surveillance system simply because the reported cases represent such a small percentage of the total number of cases occurring. Potential biases can be detected in the surveillance system by comparing surveillance data with that obtained for intensively studied groups for whom reporting is nearly complete such as in an epidemic setting. The seasonality of sporadic cases, the increased risk of illness in males, in persons over 50 or in persons with underlying disease, and the association of illness with exposure to excavation sites have all been observed in outbreaks. These risk factors for sporadic disease are therefore likely to be real.

V. INCIDENCE AND PREVALENCE

The indirect fluorescent antibody (IFA) assay, with an estimated sensitivity of 75 to 91% and specificity of 98% for LD, has two uses, clinical and epidemiologic. Specificity of the test is likely to be highest among patients with pneumonias, as the predictive value of a positive IFA result depends on the prevalence of disease being evaluated in a population under study. The test can, therefore, assist the clinician by supporting the diagnosis of LD in an individual case.

Serologic data from serosurveys of groups may allow assessment of the aggregate extent

* Relative Risk.

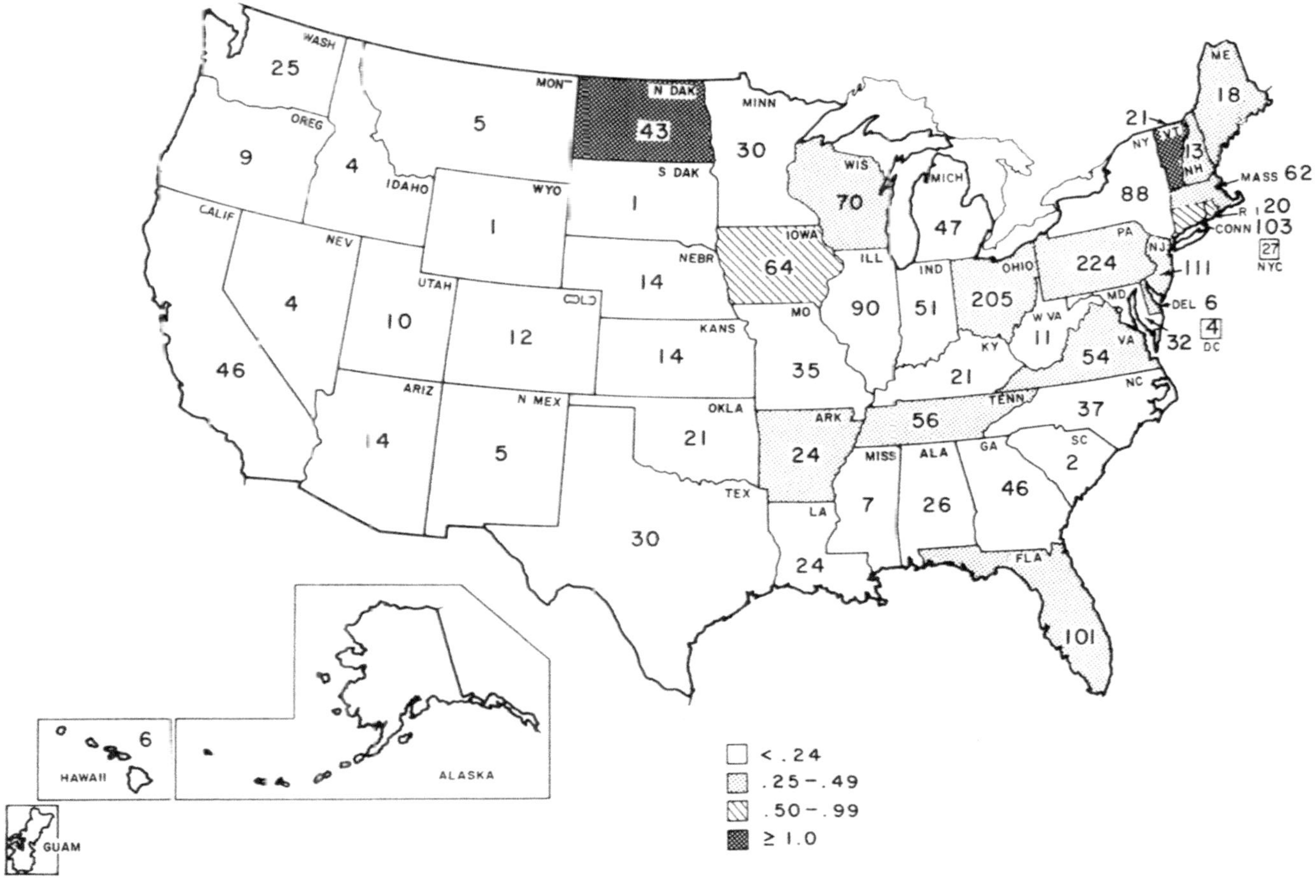

FIGURE 2. Confirmed sporadic cases and average incidence of Legionnaires' disease, by year and state, U. S., 1978 to 1981.

of past exposure if the assay is specific. However, the value of the IFA in either clinical or epidemiologic studies is also affected by the possibility of cross-reactivity. Cross-reactivity, i.e., to *Mycoplasma* and *Bacteroides fragilis* may decrease the usefulness of the IFA test results, by lowering specificity. There also, however, may be asymptomatic seroconversion to *Legionella*.

Lattimer conducted a serologic follow-up study of 31 survivors of LD during the 1976 Philadelphia outbreak.[31] He found that 94% of these Legionnaires had elevated specific IgG, IgM, or both compared with 2% of 61 Legionnaire control subjects who had not attended the convention in Philadelphia and with 3% of 29 non-Legionnaire control subjects seen at a community hospital for pneumonia. However, in 53% of the 90 Legionnaires who attended the convention and failed to develop an illness compatible with Legionnaires' disease, significant levels of IgG or IgM were present two years after the exposure. Sensitivity and specificity of the IFA assay are discussed in more detail in the chapter on serology.

Lattimer's study also suggested that elevated titers against *L. pneumophila* may persist for several years. These findings support what has been found in a follow-up study of a very small sample of cases that occurred at St. Elizabeth's Hospital in the 1965 outbreak. Of 26 persons who had serologic evidence of LD in 1965, 4 could be located. Of these 4, 3 patients continued to have diagnostic titers in 1977.

Foy[32] conducted a retrospective study of stored paired sera from 500 patients treated for pneumonia in a prepaid medical group in Seattle, to attempt to determine the absolute incidence of LD in that community from 1963 to 1975. Only 5 (1%) had a fourfold rise in antibody titers to the LD antigens, yielding an estimated incidence of disease of between 0.4 to 2.8/10,000 persons/year. However, the study population probably under sampled patients hospitalized for pneumonia. Therefore, these figures may well have underestimated the actual incidence of LD in that community.

Renner[33] attempted to establish the relative incidence of LD by testing paired sera submitted to a university laboratory between 1972 and 1977 on patients with clinical diagnosis of pneumonia due to viral, mycoplasmal, chlamydial, or unknown cause. A confirmed case was defined as one with a fourfold or greater rise in indirect FA titers to at least 1:128, and a probable case as one in which no significant titer rise occurred but in which one of the paired sera had a titer of 1:256 or more. During the 6-year period, Renner found 24 confirmed and 67 presumptive cases of LD among 586 persons with pneumonia for whom sera was submitted to the laboratory. The frequency of confirmed cases ranged from 1.1% in 1973 to 7.5% in 1976, with an overall frequency of 4.1%. This is about twice the incidence noted for a similar group of patients with pneumonia studied by McDade.[2] Renner also studied paired sera from 463 patients with pneumonia, using complement fixation antibody assays for serologic evidence of influenza A and B, adenovirus, and *Mycoplasma*, and by IFA for *Legionella* antigens. Only 22% of the paired sera gave results positive for a specific diagnosis. LD infections (4.5%) ranked third in incidence behind *M. pneumoniae* (10.2%), and influenza A (5.8%)

Legionnaires' disease has been reported in children and infants albeit infrequently.[34] To try to establish the relative frequency of *L. pneumophila* infections in children, Anderson prospectively studied 52 children less than 4 years old with acute lower respiratory tract infections.[35] Viral, mycoplasmal, and bacterial cultures, and acute and convalescent sera were obtained. In addition, annual follow-up sera was obtained and evaluated for *L. pneumophila*. Acute and convalescent sera were collected for 64 episodes of lower respiratory tract infection. Seventy-five percent of these revealed either viral or *Mycoplasma* infection. In the remaining 16 (25 %), no etiology was identified. None of the children had a fourfold or greater rise to *L. pneumophila* associated with acute illness. However, sera obtained after annual checkups revealed that 27 (52%) of the 52 children had a fourfold or greater rise in reciprocal antibody titer to $\geq$128. Therefore, although it seems that *Legionella* is not a common cause of symptomatic lower respiratory tract infections in children, exposures to the organism occurs and immunity develops, most frequently to serogroup 2.

Muldoon et al.[36] conducted a serosurvey of 35 children from 1 month to 5 years of age on whom paired sera were submitted to the Division of Virology for serologic diagnosis of suspected viral infection. Three of the 35 children had evidence of a fourfold or greater rise in reciprocal antibody titer to *Legionella* ≥128. Two of these children had roentgenographic evidence of pneumonia, one had a clinical diagnosis of bronchiolitis. These findings suggest that up to 6% of pneumonia in children may be due to *L. pneumophila*.

In several serosurveys investigators have attempted to define the level of exposure to *Legionella* within communities where outbreaks of legionellosis have occurred. A serosurvey conducted by Broome in Burlington, Vermont, revealed a range of from a high of 26% seropositivity among hospital employees, to a low of 8% among household contacts.[37] A second serosurvey was conducted in Bloomington, Indiana, in conjunction with an investigation of an outbreak of 11 cases of LD.[13] Titers of 1:128 were detected for 17% of employees of the student union. Cordes[15] investigated the prevalence of seropositivity for LD in the New York City garment district during September 1978. The rates of reciprocal titers of ≥128 for workers in Manhattan were impressive; 34.9% for garment district workers, 36.8% for workers in a control area within Manhattan, and a rate of 20.6% in a control area in the Bronx. Haley[17] conducted a serosurvey of employees at the Wadsworth VA Medical Center, a site of a continuing nosocomial outbreak. He found that 20 of 119 (17%) Medical Center employees had titers of 1:128 or greater compared with only 5 of 116 (4%) of employees in a Veteran's Administration Regional Office building located less than 0.5 km away.

Regional studies have also been conducted to determine the prevalence of seroreactivity in the absence of disease. IFA testing of single serum specimens were used to establish background prevalence of antibody to *L. pneumophila* in 1,143 persons 46 years of age and older who resided in Rochester, New York, Atlanta, Georgia, Washington, D. C., and Houston, Texas.[38] The overall prevalence of seropositivity, in this study defined as a titer of ≥64, was 1.7%.

Muldoon,[36] in an attempt to establish a baseline level seroreactivity prevalence in an adult community, tested 56 sera collected from women free of respiratory illness aged 15 to 45 seen in a prenatal clinic. At least 32% had a titer of 128 or greater on a single serum sample.

Cohen[39] attempted to ascertain the incidence of fatal nosocomial Legionnaires' disease by direct FA examination of autopsy specimens submitted over a 3-year period by hospitals participating in the National Nosocomial Infections Study (NNIS). Lung tissue was submitted on cases of culture-negative pneumonia as well as a sample of nosocomial pneumonia on which a pathogen was reported or for which no cultures were obtained. Tissues from 263 patients with nosocomial pneumonia were submitted by 40 hospitals from 24 states. Ten patients (3.8%) were identified as having had LD. Interestingly, cultures on 5 patients had led to isolation of a non-*Legionella* pathogen. During the time covered by the study, October 1974 to September 1977, 2,756 patients reported to NNIS died with nosocomial pneumonia, leading to an estimated 105 cases of fatal LD among these hospitals per year. Since NNIS hospitals account for 3.7% of hospitalized patients in the U.S., Cohen extrapolated further, for an incidence of an estimated 2,850 fatal cases from October 1974 to September 1977, or 950 cases in the country per year.

VI. EPIDEMIOLOGY OF NON-*PNEUMOPHILA LEGIONELLA*

The study of non-*pneumophila* legionellosis is still in the initial stages. Non-*pneumophila Legionella* species were associated with only 8% of the cases reported to CDC for whom etiologic information was available. Of these, *L. micdadei* was most frequently isolated, but not at a rate significantly different from that of the least frequently isolated organism, *L. dumoffi*.

Three small outbreaks of non-*pneumophila Legionella* infection have been recognized: all were associated with *L. micdadei*. One outbreak occurred in 1978 among 5 immunosuppressed patients hospitalized in Charlottesville, Va., a second and third involved 14 immunosuppressed patients and occurred in 2 Pittsburgh hospitals in 1978 and 1979 to 1980.[41-45] The age range was 30 to 80 years, and the male-to-female ratio showed male predominance in only 1 of the 3 outbreaks. Illness was associated with pneumonia in all patients, although no roentgenographic pattern was characteristic. The overall case-fatality rate was 37%. The high rate of pneumonia and death among cases may have, at least in part, been the result of a bias in case finding toward patients with pneumonia detected at autopsy. Of those patients treated with erythromycin, the case-fatality rate was 0%.[45] A sporadic case of pneumonia secondary to *L. micdadei* infection has also been reported in a 60-year-old man admitted to a hospital for elective surgery and discharged on steroids.[46]

Since 1959 when Bozeman, Humphries, and Campbell[21] isolated WIGA *(L. bozemanii)* from lung tissue of a patient who had died of bronchopneumonia, the organism has been isolated from postmortem lung specimens of an additional patient with pneumonia.[47] The patient was a 60-year-old man with a history of chronic lymphocytic leukemia who had fallen into brackish water several weeks before he developed *Legionella* infection.

Legionella dumoffii has been isolated from 2 patients with pneumonia: (1) a 60-year-old man with oat cell carcinoma, and (2) an 18-year-old woman with a history of systemic lupus erythematosis.[48,49] Both were receiving steroid therapy at the time the pneumonia developed. The organism has also been isolated from cooling tower water in New York City during an investigation of an outbreak of Legionnaires' disease.

Four cases of sporadic disease associated with *L. longbeachae* have been reported: 3 in men and 1 in a woman, ranging in age from 51 to 87 years.[50] The 3 men had a history of underlying medical problems, i.e., rheumatoid arthritis, sarcoidosis, and prostatic cancer. All developed pneumonia. Interestingly, 3 of the 4 also developed a pleural effusion.

A recently described species, *L. wadsworthii*, has also been cited as a cause of pneumonia in a patient with chronic lymphocytic leukemia.[51]

L. gormanii has been isolated from soil collected from a creek bank. It may also be associated with human disease.[52] While the organism has never been isolated from human tissue, DFA testing of formalinized lung specimens obtained at autopsy from a 69-year-old man who died of pneumonia was positive for this organism.

While it is clear that infection with the non-*pneumophila* species, in particular *L. micdadei*, can be nosocomial and opportunistic, the small number of cases recognized thus far limit further epidemiologic evaluation.

Despite the dramatic increase in our knowledge since the 1976 Philadelphia outbreak, several perplexing questions about the epidemiology of legionellosis remain. One of the major questions is the reason for the difference in the epidemiology of Pontiac fever and Legionnaires' disease, two distinct illnesses caused by the same organism. No differences can be attributed to the organism or to the circumstances of outbreaks to explain the higher attack rate, shorter incubation period, and milder clinical illness of the non pneumonic form. Second, the role of potable water in epidemic legionellosis remains unclear. Several studies have implicated potable water. In addition, the importance of the concentration of legionellae in water and soil remains unclear. Although the organism is ubiquitous, only certain locations have been associated with disease. Finally, the ecology of cooling towers warrants further definition, including studies on factors that permit dissemination of virulent organisms, and appropriate control measures.

Although many questions remain to be answered with regard to legionellosis, the data gathered to date allow us to design specific approaches for future investigations of this group of diseases.

REFERENCES

1. **Fraser, D. W., Tsai, T. R., Orenstein, W., Parkin, W. E., Beecham, H. J., Sharrar, R. G., Harris, J., Mallison, G. F., Martin, S. M., McDade, J. E., Shepard, C. C., Brachman, P. S., Field Investigation Team,** Legionnaires' disease: description of an epidemic of pneumonia, *N. Engl. J. Med.*, 297, 1189, 1977.
2. **McDade, J. E., Shepard, C. C., Fraser, D. W., Tsai, T. R., Redus, M. A., Dowdle, W. R., and the Laboratory Investigation Team,** Legionnaires' disease: isolation of a bacterium and demonstration of its role in other respiratory disease, *N. Engl. J. Med.*, 297, 1197, 1977.
3. **Osterholm, M. T., Chin, T. D. Y., Osborne, D. O., Dull, H. B., Dean, A. G., Fraser, D. W., Hayes, P. S., and Hall, W. N.,** A 1957 outbreak of Legionnaires' disease associated with a meat-packing plant, *Am. J. Epidemiol.*, 117, 60, 1983.
4. **Thacker, S. B., Bennett, J. V., Tsai, T. F., Fraser, D. W., McDade, J. E., Shepard, C. C., Williams, K. H., Jr., Stuart, W. H., Dull, H. B., and Eickhoff, T. C.,** An outbreak in 1965 of severe respiratory illness caused by the Legionnaires' disease bacterium, *J. Infect. Dis.*, 138, 512, 1978.
5. **Terranova, W., Cohen, M. L., and Fraser, D. W.,** 1974 Outbreak of Legionnaires' disease diagnosed in 1977. Clinical and epidemiological features, *Lancet*, 2, 122, 1978.
6. **Cordes , L. G., Fraser, D. W., Skaliy, P., Perlino, C. A., Elsea, W. R., Mallison, G. F., and Hayes, P. S.,** Legionnaires' disease outbreak at an Atlanta, Georgia, country club: evidence for spread from an evaporative condenser, *Am. J. Epidemiol.*, 111, 425, 1980.
7. **England, A. C., II, Goings, S. A. J., and Fraser, D. W.,** Legionnaires' disease associated with a convention in Dallas, 1978 (manuscript).
8. **Band, J. D., Laventure, M., Davis, J. P., Mallison, G. F., Skaliy, P., Hayes, P. S., Schell, W. L., Weiss, H., Greenberg, D. J., and Fraser, D. W.,** Epidemic Legionnaires' disease. Airborne transmission down a chimney, *JAMA*, 245, 2404, 1981.
9. **Dondero, T. J., Rendtorff, R. C., Mallison, G. F., Weeks, R. M., Levy, J. S., Wong, E. W., and Schaffner, W.,** An outbreak of Legionnaires' disease associated with a contaminated air-conditioning cooling tower, *N. Engl. J. Med.*, 302, 365, 1980.
10. **Cordes, L. G., Wiesenthal, A. M., Gorman, G. W., Phair, J. P., Sommers, H. M., Brown, A., Yu, V. L. , Magnussen, M. H., Meyer, R. D., Wolf, J. S., Shands, K. N., and Fraser, D. W.,** Isolation of *Legionella pneumophila* from hospital shower heads, *Ann. Intern. Med.*, 94, 195, 1981.
11. **Glick, T. H., Gregg, M. B., Berman, B., Mallison, G., Rhodes, W. W., and Kassanoff, I.,** Pontiac fever. An epidemic of unknown etiology in a health department. I. Clinical and epidemiologic aspects, *Am. J. Epidemiol.*, 107, 149, 1978.
12. **Fraser, D. W., Deubner, D. C., Hill, D. L., and Gilliam, D. K.,** Nonpneumonic, short-incubation-period legionellosis (Pontiac fever) in men who cleaned a steam turbine condenser, *Sci*, 205, 690, 1979.
13. **Politi, B. D., Fraser, D. W., Mallison, G. F., Mohatt, J. V., Morris, B. K., Patton, C. M., Feeley, J. C., Telle, R. D., and Bennett, J. V.,** A major focus of Legionnaires' disease in Bloomington, Indiana. *Ann. Intern. Med.*, 90, 587, 1979.
14. **Dondero, T. J. Jr., Clegg, H. W., Tsai, T. F., Weeks, R. M., Duncan, E., Stickler, J., Chapman, C., Mallison, G. F., Politi, B., Potter, M. E., and Schaffner, W.,** Legionnaires' disease in Kingsport, Tennessee, *Ann. Intern. Med.*, 90, 569, 1979.
15. **Cordes, L. G., Goldman, W. D., Marr, J. S., Friedman, S. M., Band, J. D., Rothschild, E. O., Kravitz, H., Feeley, J. C., Fraser, D. W., and the Field Investigation Team,** Legionnaires' disease in New York City, August-September 1978, *Bull. N. Y. Acad. Med.*, 56, 467, 1980.
16. **Beaty, H. N., Miller, A. A., Broome, C. V., Goings, S., and Phillips, C. A.,** Legionnaires' disease in Vermont, May to October 1977, *JAMA*, 240, 127, 1978.
17. **Haley, C. E., Cohen, M. L., Halter, J., and Meyer, R. D.,** Nosocomial Legionnaires' disease: a continuing common-source epidemic at Wadsworth Medical Center, *Ann. Intern. Med.*, 90, 583, 1979.
18. **Wong, E. S., Brink, E. W., Mallison, G. F., and Fraser, D. W.,** Legionnaires' disease in Connecticut, 1978 : a nosocomial outbreak associated with a contaminated water-cooling tower, in *Proc. 19th Interscience Conference Antimicrobial Agents Chemotherapy*, Boston, Abstr. No. 483, 1979.
19. **Marks, J. S., Tsai, T. F., Martone, W. J., Barton, R. C., Kennicott, J., Holtzhauer, F. J., Baird, I., Fay, D., Feeley, J. C., Mallison, G. F., Fraser, D. W., and Halpin, T. J.,** Nosocomial Legionnaires' disease in Columbus, Ohio, *Ann. Intern. Med.*, 90, 565, 1979.
20. **Tatlock, H.,** A rickettsia-like organism recovered from guinea pigs, *Proc. Soc. Exp. Biol. Med.*, 57, 95, 1944.
21. **Bozeman, F. M., Humphries, J. W., and Campbell, J. M.,** A new group of rickettsia-like agents recovered from guinea pigs, *Acta Virol. (Praha)*, 12, 87, 1968.
22. **Hebert, G. A., Thomason, B. M., Harris, P. P., Hicklin, M. D., and McKinney, R. M.,** "Pittsburgh pneumonia agent": a bacterium phenotypically similar to *Legionella pneumophila* and identical to the Tatlock bacterium, *Ann. Intern. Med.*, 92, 53, 1980.

23. **Herbert, G. A., Moss, W. C., McDougal, L. K., Bozeman, F. M., McKinney, R. M., and Brenner, D. J.,** The rickettsia-like organisms TATLOCK (1943) and HEBA (1959): bacteria phenotypically similar to but genetically distinct from *Legionella pneumophila* and the WIGA bacterium, *Ann. Intern. Med.*, 92, 45, 1980.
24. **Jackson, C. B., Crocker, T. T., and Smadel, J. E.,** Studies on two rickettsia-like agents probably isolated from guinea pigs, *Bacteriol. Proc.*, 119, 1952.
25. **England, A. C., Fraser, D. W., Plikaytis, B. D., Tsai, T. F., Storch, G., and Broome, C. V.,** Sporadic legionellosis in the United States: the first thousand cases, *Ann. Intern. Med.*, 94, 164, 1981.
26. **Fliermans, C. B., Cherry, W. B., Orrison, L. H., Smith, S. J., Tison, D. L., and Pope, D. H.,** Ecological distribution of *Legionella pneumophila, Appl. Environ. Microbiol.*, 41, 9, 1981.
27. **Tobin, J. O'H., Swann, R. A., and Bartlett, C. L. R.,** Isolation of *Legionella pneumophila* from water systems: methods and preliminary results, *Br. Med. J.*, 282, 515, 1981.
28. **Kaufmann, A. F., McDade, J. E., Patton, C. M., Bennett, J. V., Skaliy, P., Feeley, J. C., Anderson, D. C., Potter, M. E., Newhouse, V. F., Gregg, M. B., and Brachman, P. S.,** Pontiac fever: isolation of the etiologic agent *(Legionella pneumophila)* and demonstration of its mode of transmission, *Am. J. Epidemiol.*, 114, 337, 1981.
29. **Shands, K. N., Ho, J. L., Gorman, G. W., Meyer, R. D., Edelstein, P. H., Finegold, S. M., and Fraser, D. W.,** Potable water as a source of Legionnaires' disease, *Clin. Res.*, 29, 260A, 1981.
30. **Storch, G., Baine, W. B., Fraser, D. W., Broome, C. V., Clegg, H. W., II, Cohen, M. L., Goings, S. A. J., Politi, B. D., Terranova, W. A., Tsai, T. F., Plikaytis, B. D., Shepard, C. C., and Bennett, J. V.,** Sporadic community-acquired Legionnaires' disease in the United States. A case-control study, *Ann. Intern. Med.*, 90, 596, 1979.
31. **Lattimer, G. L., Rhodes, L. V., III, Salventi, J. S., Galgon, J. P., Stonebraker, V., Boley, S., and Haas, G.,** The Philadelphia epidemic of Legionnaires' disease: clinical, pulmonary and serologic findings two years later, *Ann. Intern. Med.*, 90, 522, 1979.
32. **Foy, H. M., Broome, C. V., Hayes, P. S., Allan, I., Cooney, M. K., and Tobe, R.,** Legionnaires' disease in a prepaid medical-care group in Seattle, 1963-75, *Lancet*, 1, 767, 1979.
33. **Renner, E. D., Helms, C. M., Hierholzer, W. J., Hall, N., Wong, Y. W., Viner, J. P., Johnson W., and Hausler, W. J.,** Legionnaires' disease in pneumonia patients in Iowa. A retrospective seroepidemiologic study, 1972-1977, *Ann. Intern. Med.*, 90, 603, 1979.
34. **Orme, R. L'E., Haas, L., Cruickshank, J. G., Hart, R. J. C., and Anderson, A. W.,** Legionnaires' disease in infants, *Lancet*, 2, 1027, 1980.
35. **Andersen, R. D., Lauer, B. A., Fraser, D. W., Hayes, P. S., and McIntosh, K.,** Infections with *Legionella pneumophila* in children, *J. Infect. Dis.*, 143, 386, 1981.
36. **Muldoon, R. L., Jaecker, D. L., and Kiefer, H. K.,** Legionnaires' disease in children, *Pediatrics*, 67, 329, 1981.
37. **Broome, C. V ., Goings, S. A. J., Thacker, S. B., Vogt, R. L., Beaty, H. N., Fraser, D. W., and the Field Investigation Team,** The Vermont epidemic of Legionnaires' disease, *Ann. Intern. Med.*, 90, 573, 1979.
38. **Storch, G., Hayes, P. S., Hill, D. L., and Baine, W. B.,** Prevalence of antibody to *Legionella pneumophila* in middle-aged and elderly Americans, *J. Infect. Dis.*, 140, 784, 1979.
39. **Cohen, M. L., Broome, C. V., Paris, A. L., Martin, W. T., and Allen, J. R.,** Fatal nosocomial Legionnaires' disease: clinical and epidemiologic characteristics, *Ann. Intern. Med.*, 90, 611, 1979.
40. **Rogers, B. H., Donowitz, G. R., Walker, G. K., Harding, S. A., and Sande, M. A.,** Opportunistic pneumonia. A clinicopathological study of five cases caused by an unidentified acid-fast bacterium, *N. Engl. J. Med.*, 301, 959, 1979.
41. **Pasculle, A. W., Myerowitz, R. L., and Rinaldo, C. R., Jr.,** New bacterial agent of pneumonia isolated from renal-transplant recipients, *Lancet*, 2, 58, 1979.
42. **Myerowitz, R. L., Pasculle, A. W., Dowling, J. N., Pazin, G. J., Puerzer, M., Yee, R. B., Rinaldo, C. R., Jr., and Hakala, T. R.,** Opportunistic lung infections due to "Pittsburgh pneumonia agent", *N. Engl. J. Med.*, 301, 953, 1979.
43. **Pasculle, A. W., Feeley, J. C., Gibson, R. J., Cordes, L. G., Myerowitz, R. L., Patton, C. M., Gorman, G. W., Carmack, C. L., Ezzell, J. W., and Dowling, J. N.,** Pittsburgh pneumonia agent: direct isolation from human lung tissue, *J. Infect. Dis.*, 141, 727, 1980.
44. **Thomason, B. M., Ewing, E. P., Hicklin, M. D., Harding, S. A., and Downowitz, G. R.,** TATLOCK bacteremia (Pittsburgh pneumonia agent) presumptively identified in five cases of pneumonia, *Ann. Intern. Med.*, 92, 510, 1980.
45. **Wing, E. J., Schafer, F. J., and Pasculle, A. W.,** Successful treatment of *Legionella micdadei* (Pittsburg pneumonia agent) pneumonia with erythromycin, *Am. J. Med.*, 71, 836, 1981.
46. **Donegan, E. A., Deal, M. M., Melanephy, M. C., Cordes, L. G., Moss, C. W., Pons, V. G., and Brooks, G. F.,** Primary isolation of a new strain of Tatlock/Pittsburg pneumonia agent *(Legionella micdadei), West. J. Med.*, 134, 384, 1981.

47. **Thomason, B. M., Harris, P. P., Hicklin, M. D., Blackmon, J. A., Moss, C. W., and Matthews, F.,** A *Legionella*-like bacterium related to WIGA in a fatal case of pneumonia, *Ann. Intern. Med.*, 91, 673, 1979.
48. **Lewallen, K. R., McKinney, R. M., Brenner, D. J., Moss, C. W., Dail, D. H., Thomason, B. M., and Bright, R. A.,** A newly identified bacteria phenotypically resembling, but genetically distinct from *Legionella pneumophila:* an isolate in a case of pneumonia, *Ann. Intern. Med.*, 91, 831, 1979.
49. **Harris, P. P., Aufdemorte, T., Ewing, E. P., Jr., Johnson, J. E., and Tio, F. O.,** Fluorescent-antibody detection of *Legionella dumoffi* in a fatal case of pneumonia, *J. Clin. Microbiol.*, 13, 778, 1981.
50. **McKinney, R. M., Porschen, R. K., Edelstein, P. H., Bissett, M. L., Harris, P. P., Bondell, S. P., Steigerwalt, A. G., Weaver, R. E., Ein, M. E., Lindquist, D. S., Kops, R. S., and Brenner, D. J.,** *Legionella longbeachae* species nova, another etiologic agent of human pneumonia, *Ann. Intern. Med.*, 94, 739, 1981.
51. **Edelstein, P. H., Brenner, D. J., Moss, C. W., Steigerwalt, A. G., Francis, E. M., and George, W. L.,** *Legionella wadsworthii* species nova: a cause of human pneumonia, *Ann. Intern. Med.*, 97, 809 , 1982.
52. **Morris, G. K., Steigerwalt, A., Feeley, J. C., Wong, E. S., Martin, W. T., Patton, C. M., and Brenner, D. J.,** *Legionella gormanii* sp. nov., *J. Clin. Microbiol.*, 12, 718, 1980.
53. **Gregg, M.,** personal communication.
54. **Neill, M.,** unpublished results.

Chapter 6

ECOLOGICAL NICHE OF *LEGIONELLA PNEUMOPHILA*

Carl B. Fliermans

TABLE OF CONTENTS

I. INTRODUCTION

Legionella is a most fascinating bacterium. It has properties and capabilities that if we as scientists have ears to hear and eyes to see, *Legionella* will indeed tell us some fascinating stories about the world in which we live and the responsibilities we have to this creation. The scope of this chapter is directed towards the ecological niches, relationships, and controls of *Legionella* derived from environmental sources. Only which clinical cases and studies related directly to the ecological understanding of the bacterium will they be discussed.

Not until these recent times has such a successful marriage between medical microbiology and microbial ecology been consummated. The format for such a marriage was established with Berkeley's work in 1845 with Irish Potato Blight and Robert Koch's work in 1876 on anthrax, a disease of cattle caused by *Bacillus anthracis*. The use of Koch's Postulates developed in the study of medical microbiology is equally applicable to the study of microbial ecology. Koch's Postulates in microbial ecology would be defined as the following:

1. The selected microorganism should always be found in the habitat that is responding in a characteristic manner, and should not be present in habitats that do not respond in such a manner.
2. The organism must be isolated and cultivated in pure culture away from the environment.
3. The isolated culture, when inoculated back into the environment or a subsample of that environment, should produce the characteristics of the habitat containing the microorganism.
4. The microorganism must then be reisolated from the inoculated habitat, cultured in the laboratory, and be the same microorganism as the original.

Koch's contributions set the stage for subsequent successful ecological studies over the past century, and it is from this heritage that the ecological research on *Legionella* draws its breath.

Several recent reviews[1-3] have covered the clinical, epidemiological, and technical aspects of the disease with little attention given to *Legionella*'s ecological role. This review seeks to separate the ecological parameters associated with *Legionella* that are often incorporated into the medical literature as well as to highlight specific ecological studies. The data from published results through July 1982 are pulled together in order to weave a story that allows the deaf to hear and the blind to see. A series of ecological studies will demonstrate the niches of *Legionella*, the ecological parameters that allow the bacterium to survive, grow and to be disseminated. Relationships among given habitats will be explored along with biological relationships within a given habitat. The understandings gleaned from the environmental considerations will provide the focus for new research directions in order to better predict and control the exposure of *Legionella* to a susceptible human population. Thus through the acquisition of knowledge, the goal of understanding more clearly is viewed and wisdom is enhanced.

II. TECHNIQUES OF ISOLATION FROM ENVIRONMENTAL SOURCES

Within the aquatic environment a large and diverse number of microorganisms exist. Based on density enumeration, the bacterial community generally composes the dominant form. Although the diversity of the bacterial population may be very large, specific bacterial components such as *Legionella* may constitute a very small portion of the total population. In order to determine the presence and density of selected bacteria in a water sample, the sample must be concentrated in such a manner that the bacteria of choice are readily separated and identified from the surrounding milieu. Because *Legionella* comprises a rather minor portion of the total bacterial population in a lake or river system, concentration of the bacterium is necessary.

Initially, Fliermans et al.[4,5] used continous flow centrifugation to concentrate 500-fold the total flora and fauna of the water samples. The technique utilizes a Sent-Gyorgyi and Blum 8-tube continuous flow centrifuge system (Sorvall, Inc., Norwalk, Conn.) attached to a high speed centrifuge (SS-3, Sorvall, Inc.). Large volumes of water can be passed through the centrifuge in a rather short period of time. Generally, 8 to 20 ℓ of water are processed at a rate of 300 mℓ/min at 10,000 rpm. The recovery rate for Legionella seeded water samples has been consistently around 80%. The recovery rate will vary somewhat with the type of water sample, since the attachment of *Legionella* to other organisms in the aquatic community will effect the size distribution and thus the settling velocity.

The advantage of the technique is clearly its ability to concentrate large volumes of water in a relatively short period of time. In the study of waters from finished drinking water or very clean underground wells, the use of large volumes is necessary since the total *Legionella* population is generally quite low.[4] Continuous flow centrifugation has been used in field sampling in conjunction with the use of a mobile laboratory as well as bringing the samples to a centrally located laboratory for processing.[6]

The major disadvantage of the continuous flow centrifugation technique is that the expense of the initial setup is sizeable, in that a single setup may run $6000 to $7000. Once the sample is pelleted into the metal centrifuge tubes, the eight tubes are washed free of the material and the volume is adjusted to a 500-fold final concentration.[5] This level of concentration was established as a practical working definition and allows the detection of 9100 *Legionella* per liter of unconcentrated water.[5] The washing of the pellet from the centrifuge tubes constitutes an additional handling procedure and is a source of error for even the most cautious, since material in some aquatic habitats remains flocculent and will prevent the complete recovery of any *Legionella* associated with such material.

Orrison et al.[7] have demonstrated the potential of using membrane filtration for the recovery of *Legionella* from cooling tower water samples. Samples, 1 to 2 ℓ were collected and passed through a 0.45 or 0.65 nm Millipore filter (Millipore Corp., Bedford, Mass.) under positive nitrogen pressure. The filter was blended and centrifuged at 9000 × g for 1 hr and brought to a final volume of 5.0 mℓ for a concentration factor of 200 to 400 times, depending on the initial volume filtered. Reported recovery efficiencies ranging from 9 to 78% of the *Legionella* seeded into the water sample were typical. Studies by Braun[8] indicated that recovery rates using Millipore 0.45 nm filters under vacuum were fairly consistent and reproducible with rates of 40 to 45% using pure cultures. The differences in the recovery rates are not clear. The use of Nuclcopore filters in place of Millipore might be expected to increase the rates of recovery since bacteria can easily be trapped within the matrix of a Millipore filter. It must be stressed that the water samples need to be of relatively small volumes or extremely clean to allow large volumes to pass through the 0.4 nm pore size that is necessary to retain the *Legionella.*

The obvious advantage of the filtration technique is that virtually all laboratories are equipped in such a way as to be able to filter some volume of water in a reliable manner. The equipment is relatively inexpensive and straightforward. Orrison et al.[9] have used positive pressure filtration, which appears to be quite adequate. Although the amount of pressure used for the filtration process was not indicated, we have used up to 80 psi with a Nucleopore apparatus without any indication of harm to the bacterium. The positive pressure allows for greater volumes of water to be filtered because vacuum filtration achieves only approximately 15 psi. There is, however, a second step that is used in the filtration process that leads to additional handling and possible source of error.[9] Once the water sample has been filtered, the entire filter is placed in 50 mℓ of filtrate and blended. The sample is then centrifuged and the sample decanted and concentrated to the desired level. If care were not

Table 1
RECOVERY RATES OF *L. PNEUMOPHILA* SEROGROUPS 1 TO 4 FROM WATER SAMPLES TAKEN FROM PAR POND

Comparison is made between continuous flow centrifugation and positive pressure filtration at 80 psi through a 0.40 nm Nucleopore filter (142 nm)

Sample	Vortex (min)	*Legionella/ℓ*	% Recovery
Centrifuge	0	$3.28 \times 10^5 + 0.3^a$	100.0
Filtration	0.5	$2.00 \times 10^5 \pm 0.3$	61.0
Filtration	1.0	$3.09 \times 10^5 \pm 0.3$	94.3
Filtration	2.0	$2.73 \times 10^5 \pm 0.3$	83.2
Filtration	4.0	$3.91 \times 10^5 \pm 0.3$	119.3
Filtration	10.0	$3.46 \times 10^5 \pm 0.3$	105.4
Filtration	20.0	$3.19 \times 10^5 \pm 0.3$	97.1
Filtration	30.0	$3.46 \times 10^5 \pm 0.3$	105.4

[a] ± 1 Standard error.

taken in this additional step of centrifugation, the resulting densities could be radically altered by the slight change in volumes upon dilution. This second handling step can be avoided through the use of polycarbonate filters.

The data in Table 1 demonstrate that large Nucleopore filters (142 mm) are easily backflushed to remove the *Legionella* trapped on the filter. The data compare the recovery of split water samples from a cooling lake in South Carolina called Par Pond.[10-12] The levels of recovery by filtration with Nucleopore filters are equal to or better than those for the same sample using continuous flow centrifugation. *Legionella* appears to be readily washed from the filter, and mixing times up to 30 min do not appear to have an unfavorable effect on the recovery of *Legionella* when compared to the continous flow centrifugation procedure previously described.

Along with the advantages of the filtration techniques comes the disadvantages of limited volumes as previously mentioned. Many of the present studies in the U.S. are centering around potable water, finished water, drinking water, or ground water sampling. These systems are relatively "clean" in that they generally contain very little sediments. Thus the prospect for filtration is very much enhanced. The hooker in the logic is that relatively clean systems may have low levels of *Legionella* and thus greater volumes of water are needed to concentrate the organism. With the above caveat I believe the technique has wide applicability for aquatic sampling and should be employed wherever the situation lends itself.

The assessment of the recovery rate of *Legionella* from these various methodological procedures is rather subjective. This is because the organisms that one seeds into the natural sample are organisms taken from enriched media and are not adapted to the environment in which they are tested. Because *Legionella* is associated with aquatic environments, a wide range of interactions between *Legionella* and other aquatic organisms exists. Such interactions will cause density variability and the patchiness of the *Legionella* will be profound. These associations do not occur in a pure culture and thus the recovery rate achieved by a designated technique will not be directly related to the real world where *Legionella* exists.

Although *Legionella* has been shown to be in rather low concentrations in the natural aquatic environment,[4] isolates have been obtained without concentration techniques in both natural and finished water systems.[13-18] Some of these techniques are discussed below with reference to sampling in the natural ecological habitats where the organism exists. The medium of choice remains the charcoal-yeast extract (CYE) medium that was developed by Feeley et al.[18] Once isolated all known strains and species of *Legionella* will grow on CYE. The medium however is not selective, so that any enrichment that occurs is subjected to the competition of other microorganisms. Such competition occurs in any environmental sample. Various media have been developed that are designed to be both an enrichment as well as a differential media.

Thorpe and Miller[19] designed a procedure to take advantage of the fact that *Legionella* was resistant to a large number of antibiotics, while the natural bacterial population may be more susceptible. The water sample was exposed to a nutrient broth containing agents that were static to *Legionella* while being cidal to the contaminating microbial community. The data indicated that the technique was indeed useful when *Legionella* cultures were seeded into the experiments but positive results from nonseeded water samples were not obtainable. A similar type of negative enrichment has been attempted using the thermophilic nature of *Legionella* as a possible enrichment technique; however, the isolation of *Legionella* has not been satisfactory using BCYE agar, selected antibiotics, and 45°C.[132]

Pasculle et al.[20] have reported the use of a buffered CYE agar medium for the isolation of *Legionella* from the lung tissue from infected humans. This medium, while a benefit in the growth of the organism in the laboratory, is not particularly useful as a primary isolation media for environmental isolates of *Legionella* nor was it designed as such. Wadowsky and Yee[17] have developed a medium based on the BCYE media with the addition of glycine at a final concentration of 0.3%, along with selected antibiotics to obtain the glycine-vancomycin-polymyxin B agar. The media is somewhat selective, although in our laboratory other bacteria besides *Legionella* readily grow on the medium when samples are taken from environmental sources such as lakes and rivers. Additionally, this medium appears to be somewhat inhibitory to selected strains of *L. dumoffii* and *L. micdadei*.[17]

Vickers et al.[21] described a BCYE-based medium containing bromocresol purple and bromothymol blue for the differentiation of the various members of the family Legionellaceae. The medium was designed for the separation of pure cultures of *Legionella* rather than for the purpose of isolating *Legionella* directly from the natural environment. In pure culture situations the medium appears to have applicability, however, the tested cultures are at best presumptive for the various species of *Legionella* and must be confirmed by subsequent biochemical, serological, and genetic techniques.

Recent studies by Bopp et al.[13] have suggested that the survivability of *Legionella* at low pH values may be useful in the recovery of *Legionella* from environmental samples. The technique employs the standard CYE medium to which selected antibiotics have been added. The water sample is pretreated by exposure to pH 2.2 buffer for periods of time up to 60 min. Generally for the purpose of isloation the acid exposure was 5 min. The procedure has been effective in a number of researchers' hands but has its problems. The technique is not applicable to environmental samples because a wide variety of bacteria are capable of resisting short acidic treatments and cause contamination of the medium.[22-25] The use of the acid treatment on water samples known to contain established densities of *Legionella* has caused a loss in densities from 3 to 5 logs due to the acid treatment.[133]

The use of egg yolk embryos is an established technique for isolating a wide variety of pathogenic organisms from tissues or for routine passage of laboratory strains of pathogenic organisms.[27] Initial studies[29] indicated that *Legionella* lost its virulence for either guinea pigs or embryonated eggs within ten passages on Mueller-Hinton Agar or Feeley-Gorman agar. Embryonated eggs and guinea pigs have been used to maintain the virulence of *Legionella* within the laboratory.[29] The use of embryonated eggs is not an appropriate technique for either isolating or for enriching *Legionella* from the natural environment. This is due primarily to the extensive bacteriological population that exists in the natural habitat and the abundance of organisms capable of using the embryonated egg nutrients. Further, the slow growing *Legionella* with doubling times of approximately 100 min[30] in the laboratory is no match for the other aquatic organisms that have generation times much faster and are initially in greater concentration.[12,31,32]

An overriding ecological problem with any of the above procedures is that they are basically indirect. That is, any manipulations in culturing the organism, subjects the organism to the limitations of the technique. Indirect techniques attempt to obtain information about the density and the *in situ* physiology of *Legionella* in its natural habitat via indirect means. Only under very rare circumstances, i.e., in extreme habitats,[23-26] can such indirect procedures be used with any confidence, and then the data are not always clear. The growth of *Legionella* adapted to an aquatic habitat with a myriad of interactions may not have any relation to an agar plate. The use of laboratory media to selectively enrich and differentiate *Legionella* for the purpose of evaluating the habitat from which the sample came misses the point of ecological studies. Ecological techniques should be designed to study the organism in its natural habitat, *in situ,* with the minimal disturbance so that the maximum can be learned both about the habitat and the organism. Thus, learning *Legionella* ecology through indirect laboratory manipulations is tenuous at best.

From the available data it appears that a selective, differential enrichment medium is still to be produced which allows *Legionella* to express all of its characteristics of growth and survival at elevated temperatures, low oxygen tensions, and wide ranges of pH values. The organism is certainly capable of existing in a wide variety of habitats[4] as well as growing under experimental conditions in the laboratory that are not produced by any of the presently available media.[26,33]

The use of guinea pigs as the animal host for *Legionella* has been the most widely used selective "medium" for the isolation of *Legionella* from natural habitats. Amazingly, guinea pigs are relatively resistant to most other bacteria in the aquatic habitat but relatively susceptible to the *Legionella* bacterium.[27] Our initial studies[4] on water samples from a cooling lake made use of a pet guinea pig from our children. There are certainly numerous reasons to replace the use of guinea pigs if at all possible. The expense and upkeep are but minor considerations. The isolation of *Legionella* from necropsied tissue is dependent on the host as well as the infectivity of the organisms injected. Each sample requires more than one guinea pig for reproducible results.[34,35] Samples collected from a new habitat should be always injected into guinea pigs to establish the presence of the *Legionella* rather than using the presumptive evidence from fluorescent antibodies alone. Once the presence of *Legionella* has been repeatedly confirmed within a habitat, then the use of the direct fluorescent antibody technique is relied upon with intermittent guinea pig inoculations.

The direct fluorescent antibody technique (DFA) is the most widely used technique for the screening of water samples for the presumptive presence of *Legionella*.[4,5,6,8,34,36] The technique has been shown to be sensitive and accurate for use in aquatic systems,[4,37] but must be utilized in conjunction with isolation techniques. Over 100 cultures of *Legionella* have been isolated from natural and man-made aquatic habitats. These isolates were obtained as previously described[4] using the DFA initially to screen the concentrated water samples followed by injection into guinea pigs. Upon necropsy, the interperitoneal fluid, spleen, and liver were stained with serospecific fluorescent antibodies and then the specimens were plated onto BCYE for subsequent isolation. The colonial morphology was determined and repeated isolates were again stained with DFA before subjection to fatty acid profile and DNA hybridization techniques for confirmation.[38,39]

It should be stressed that initial screening procedures for *Legionella* must always be confirmed through culture isolation, guinea pig analyses, and the subsequent routine biochemical testing.

The advantages of the DFA technique are far outweighed by any disadvantage that may be posed. The advantages will be discussed briefly from the ecological viewpoint. Presently, there is no technique known to microbial ecologists other than DFA in which a particular bacterium can be viewed directly microscopically and identified without selectively enriching the organism away from its *in situ* habitat. This point in itself is most important. The use of indirect culturing techniques is extremely necessary in order that one can study the organism in the laboratory under defined culture parameters, but such techniques do not tell one anything about the nature of the bacterium in the environment in which it lives. Once culturing of the bacterium occurs, the relationship to the real world is lost. These points are not new, but have been espoused over the years by microbiologists concerned with the difficult task of studying bacteria in their environment. The fluorescent antibody technique is extremely useful in evaluating the density, serology, and distribution of selected organisms under *in situ* conditions. The DFA technique has its roots in medical microbiology and immunology[40] but has been successfully used in the application of pathogenic and nonpathogenic organisms in aquatic and terrestrial habitats.[41-50]

As with any technique that exists in the realm of microbial ecology, there is no panacea. Disadvantages occur with every technique and the use of the DFA technique is no different. There is initially the sophistication of equipment that is required to perform the routine microscopic analyses. The set up for microscopy has been described by a number of researchers and is outlined in a paper by Fliermans and Wilde.[49,51] The cost of the set up can run anywhere from $16,000 to $75,000 depending on the various attachments that are required with the microscope. Our laboratory has over the years used primarily Zeiss equipment and found it to be most satisfactory. The versatility of the instrumentation has been the greatest plus for the Zeiss equipment. An equally viable alternative is the American Optical Scope. The resolution is sharp and the objectives look quite comparable. The cost of such a scope was at one time below that of Zeiss and may well be worth the investigation before investment. The microscope is the one piece of equipment that is vital for the use of the DFA technique. Every microbial ecology laboratory should be equipped with a high quality light microscope. It is the backbone of the laboratory, giving versatility and a wide variety of techniques.

The preparation of the antibodies for the DFA technique are a study in their own right and will not be discussed in length at this point. Over the past 10 years I have worked with a large number of ecological organisms or have been associated with laboratories that have, and none of the preparations are as selective nor as specific as the ones that are used for the study of *Legionella*. The crux of the matter is whether the antibodies for *Legionella* are specific enough to warrant their use in the natural habitats. As described previously,[4,52,53] cross reaction problems are minor and indeed the burden of proof rests on those that claim substantial cross reactions occur in the natural habitat. In our laboratory those proofs have not materialized and the technique is sound.

Figure 1 demonstrates in a rather simplistic but straightforward manner the technique that we employ for looking at the various habitats for *Legionella*. The organism is first isolated from the habitat which one would eventually like to study. Obviously with *Legionella* the initial habitat was that of a patient. Once *Legionella* is removed from the habitat, it is cultivated and isolated into pure culture, and grown to an appropriate density for injection into rabbits. The techniques used for antigen injections vary greatly, but generally the rabbits produce high quality antibodies which can be harvested by conservative methods so that the rabbit can be repeatedly used for the same antigens. The serum is harvested, purified, and coupled to a fluorescent compound, generally fluorescein isothiocyanate (FITC). The product is then termed the fluorescent antibody and is tested for specificity, proper dilution, cross reaction, and purity.[50]

To this point in the technique essentially no microbial ecological work has taken place. It is merely the preparation of the system for subsequent usage. Once the fluorescent antibody has been established, the field work begins.

Samples from the various habitats are collected and the type of sample designates the

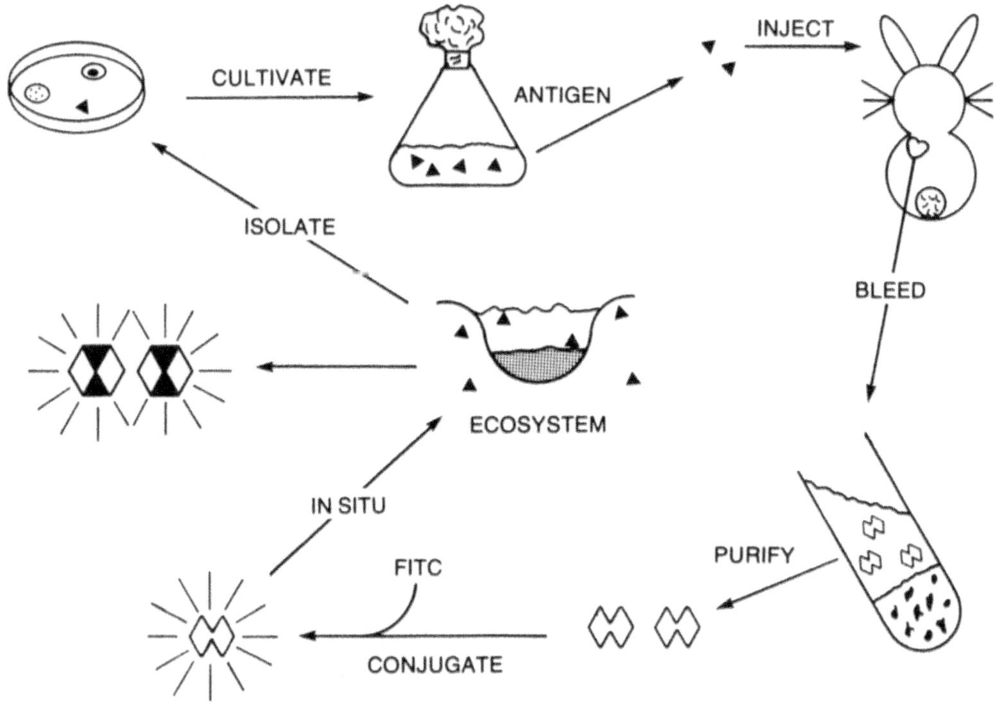

FIGURE 1. Diagram of the direct fluorescent antibody (DFA) technique for environmental systems.

methodology necessary to process the sample. The methodological aspects for the ecological study of *Legionella* from the various habitats will be considered separately. Once the sample is appropriately taken and processed, the microscopic visualization begins.

Our laboratory has chosen to maintain a microscope with the appropriate filtering combinations so that both fluorescence from the DFA technique can be observed as well as the natural fluorescence from the chlorophyll-containing organisms such as the various algal species.[49,51] The use of the DFA technique along with the necessary band-pass light filters for chlorophyll visualization allows the simultaneous viewing of juxtaposition relationships between *Legionella* and algae that would be missed by conventional fluorescence. The various excitation and emission spectra lets one distinguish the various types of chlorophyll and allows one to assess the functional status of the observed algae.[54] Thus the probable role of the algae present in the sample can be assessed.

The use of a microscope slide series permits natural populations of bacteria and algae to develop within the natural habitat.[43,55] The slides were subsequently processed for selected organisms using DFA or for total bacterial populations using a nonselective stain.[43] Using the slide series technique for a variety of bacteria, it appears that the largest concentration of aquatic bacteria are generally associated at the sediment-water interface.

Legionella differs somewhat from the normal aquatic bacterial fauna in freshwater lakes in that *Legionella* does not appear to be associated with the sediment interface area of any lakes that we have extensively studied. The data in Table 2 demonstrate that *Legionella* is well distributed throughout the upper aquatic layers of Par Pond. This is true even during time of isothermal conditions from October to May. *Legionella* does not concentrate in the sediments nor in the deeper portions of the lake below the photic zone. During the natural cycling of Par Pond, a monomictic lake, the bottom waters completely mix with the top waters.[56] The *Legionella* profiles are equally well distributed in the water column as the data in Table 3 demonstrate.

Because *Legionella* or any other bacterium that remains intact will be detected by the

Table 2.
DISTRIBUTION OF *L. PNEUMOPHILA* SEROGROUPS 1 TO 4 IN THE WATER COLUMN OF PAR POND, FEBRUARY 1980

Depth (m)	Temp. (°C)	Cond.	pH	D.O. (ppm)	Densities/ℓ Knoxville	Togus	Bloomington	Los Angeles	Isolate
Surface	11.2	64	7.0	10.6	7.4×10^5	1.1×10^5	BD	BD	K,T
1	11.2	64	7.0	10.6	1.0×10^6	5.4×10^4	BD	BD	K
2	11.2	64	7.1	10.6	6.6×10^5	1.1×10^5	5.5×10^4	BD	K
3	11.2	64	7.1	10.8	1.2×10^6	5.4×10^4	5.4×10^4	BD	K
4	11.2	64	7.1	10.8	3.8×10^5	BD	5.4×10^4	BD	K
5	11.0	64	7.1	10.8	8.2×10^5	5.4×10^4	BD	BD	K
6	11.0	64	7.1	10.8	3.5×10^5	9.1×10^3	9.1×10^3	9.1×10^3	Neg
7	11.0	64	7.1	11.4	6.6×10^5	9.1×10^3	9.1×10^3	9.1×10^3	K
8	11.0	64	7.1	11.4	1.9×10^5	9.1×10^3	BD	BD	Neg
9	11.0	64	7.1	11.4	3.2×10^5	BD	BD	BD	K
10	11.0	64	7.1	11.4	3.0×10^5	BD	BD	BD	NT
11	11.0	64	7.1	11.4	1.6×10^5	BD	BD	BD	Neg
12	11.0	64	7.1	11.4	NT	BD	BD	BD	Neg
13	11.0	64	7.1	11.4	3.8×10^5	BD	BD	BD	K
14	11.0	64	7.2	11.4	5.7×10^5	BD	BD	BD	Neg
15	10.5	66	7.2	11.6	5.7×10^5	BD	BD	7.2×10^4	Neg
16	10.5	66	7.2	11.6	8.4×10^5	9.1×10^3	BD	BD	K
0—5	NT	NT	NT	NT	4.1×10^4	BD	BD	2.7×10^4	Neg
5—10	NT	NT	NT	NT	1.4×10^4	BD	BD	2.1×10^4	Neg
10—15	NT	NT	NT	NT	4.1×10^4	BD	BD	6.9×10^3	Neg
15—20	NT	NT	NT	NT	BD	BD	BD	BD	Neg

NT = Not Tested.
BD = Below Detection ($<9.1 \times 10^3$).

Table 3
DISTRIBUTION OF *L. PNEUMOPHILA* SEROGROUPS 1 TO 4 IN WATER COLUMN OF PAR POND, AUGUST, 1980

Depth (cm)	Temp. °C	Cond.	pH	D.O.	Densities Knoxville	Togus	Bloomington	Los Angeles	Isolate
Surface	34.5	63	7.25	7.7	3.0×10^5	BD	BD	BD	K
1	32.0	64	7.75	7.7	2.7×10^5	BD	BD	BD	K
2	32.0	62	7.60	7.6	1.1×10^5	BD	BD	BD	NT
3	31.5	62	7.30	7.5	1.1×10^5	BD	2.7×10^4	BD	K
4	31.5	62	6.50	7.3	2.7×10^5	BD	BD	7.1×10^3	LA
5	30.5	62	5.80	6.8	5.7×10^5	BD	BD	BD	K
6	30.0	70	5.80	0.8	8.2×10^4	BD	BD	BD	Neg
7	25.0	80	6.20	0.1	4.6×10^6	BD	BD	BD	K
8	22.5	82	6.20	<0.1	1.7×10^6	BD	BD	BD	K
9	21.0	82	6.30	<0.1	1.3×10^5	BD	BD	BD	Neg
10	21.0	84	6.30	<0.1	3.3×10^5	BD	BD	7.6×10^6	K,LA
11	19.5	86	6.30	<0.1	7.6×10^5	BD	BD	5.5×10^4	NT
12	19.2	86	6.30	<0.1	1.3×10^6	BD	BD	BD	K
13	18.5	86	6.30	<0.1	2.5×10^5	BD	BD	3.3×10^5	LA
14	18.0	90	6.30	<0.1	2.3×10^6	BD	BD	BD	K
15	17.5	92	6.30	<0.1	1.2×10^6	BD	BD	BD	K
16	17.0	94	6.20	<0.1	1.4×10^6	BD	BD	BD	NT

NT = Not Tested.
BD = Below Detection (9.1×10^3).

DFA technique regardless of whether it is alive or dead, it becomes necessary to establish techniques whereby the difference between living and dead bacteria can be assessed. In the natural environment nonliving bacteria that are recently dead are probably very good energy and nutrient sources for other organisms that can use bacteria. In environments such as cooling towers or extreme habitats where predators may be excluded from the habitat, such probabilities do not hold. Thus it remains necessary to determine the percentage of *Legionella* that are viable in a given habitat that may have a restricted predation flora.

The technique using the tetrazolium dye INT (2-*p*-iodophenyl-3-*P*-nitro-phenyl-5-phenyl tetrazolium chloride) in combination with the DFA was developed in my laboratory to answer just this question.[30] The technique makes use of the fact that *Legionella* has an electron transport system as do most all respiring bacteria.[56,57] The oxidized form of the dye is placed in the water sample and it is taken up by actively respiring bacterial and algal cells of all types and deposited inside the cells. The oxidized form of the dye acts at the ubiquinone-cytochrome-b complex level in the electron transport system to form a reduced INT compound, formazan.[57,58] Provided the electron transport system is active in prokaryotic cells, the crystal of formazan will be deposited within the cell and can then be detected by bright field microscopy. The incorporation of the INT is not for any one type or kind of bacteria but for the various types of organisms present in the ecosystem. Studies have shown that the INT incorporation is virtually complete after 1 hr of incorporation.[30] Simultaneous viewing of the INT-formazan crystals under transmitted bright field microscopy and the specifically stained fluorescing *Legionella* by epifluorescence microscopy allows one to observe the activity of specific cells. Such methodology allows one to determine under *in situ* conditions whether the conditions in which the *Legionella* are found are conductive to their activity or detrimental. This technique is important for evaluating the effectiveness of various biocidal treatments under real world conditions like cooling towers.

The measure of activity is not a measure of virulence. We understand this from the fact that culturally grown *Legionella* can have a variety of virulence states depending on the media in which it has been grown and the number of passages that it has gone through.[29,59] If, however, the organisms in a given water sample under the environmental conditions imposed are all in the same state of virulence and activity, then the measure of activity could be a measure of virulence. To date that is merely a hypothesis and no data are available that says such is the case.

The technique of DFA and INT has been used with a number of organisms and has been used in the field extensively with *Legionella*. Monthly samples taken from the Par Pond System have been evaluated for INT activity levels of *Legionella* with regard to water column position and seasonality (Table 4). As there appears to be a seasonality to the infection levels of *Legionella* for guinea pigs[4] there appears to be greater activity of *Legionella* in the upper waters. Likewise, a seasonability of electron transport activity may occur with *Legionella* (Table 5).

The technique has provided physiological information *in situ* that would not have been obtained by any other technique.[30] Studies in the field as well as in pure culture using temperature transfer experiments have demonstrated that *Legionella* does indeed have an active electron transport system that is active over a wide range of temperatures. The INT studies support and help confirm the fact that *Legionella* is related to thermophilic bacteria in its ability to survive high temperatures and to remain active over a wide temperature range. It is significant that the highest temperature habitat from which *Legionella* has been isolated is 63°C[30] and corresponds very nicely to the upper temperature limit at which the electron transport system of *Legionella* functions as measured by the INT technique. Secondly, the isolates obtained from the field and those from clinical specimens have the same response with regard to their electron transport activity under temperature transfer experiments.[30]

Table 4
ELECTRON TRANSPORT ACTIVITY OF *L. PNEUMOPHILA* SEROGROUP 1 IN THE WATER COLUMN COLLECTED FROM AN AMBIENT TEMPERATURE STATION IN PAR POND, APRIL, 1980

Depth (m)	Temp (°C)	Number of L. pneumophila/ℓ	% Active
Surface	20.3	3.3×10^5	8.3
1	20.3	2.5×10^6	37.0
2	20.3	6.8×10^5	24.0
3	20.0	1.5×10^6	NT
4	20.0	1.3×10^6	34.8
5	20.0	5.7×10^5	NT
6	19.5	5.2×10^5	36.8
7	19.5	1.5×10^6	NT
8	19.5	3.3×10^5	33.0
9	18.5	9.6×10^5	20.8
10	18.0	2.3×10^5	16.0
11	16.5	1.3×10^6	NT
12	16.0	9.0×10^5	14.2
13	14.5	1.9×10^6	NT
14	14.0	1.9×10^5	0
15	13.5	1.8×10^6	NT
16	13.5	5.4×10^5	0

NT = Not Tested

Table 5
ELECTRON TRANSPORT ACTIVITY OF *L. PNEUMOPHILA* SEROGROUP 1 IN WATER SAMPLES COLLECTED FROM AMBIENT TEMPERATURE HABITAT OF PAR POND, SEASONALLY

Date Collected	Number of water samples measured	% Int. positive
4/80	25	30.0
5/80	10	20.0
6/80	25	28.0
7/80	25	43.4
8/80	10	43.1
9/80	25	38.6
10/80	25	16.1
11/80	25	23.8
12/80	12	41.5
1/81	25	50.0
2/82	25	41.2
3/82	10	24.0

Orrison et al.[9] have demonstrated that the environmental isolates of *Legionella pneumophila* are essentially identical with the clinical isolates in all regards. Thus there is no reason at present to separate the *Legionella* from natural habitats as opposed to those from clinical specimens. They are the same organism, capable of the same metabolic and physiological processes, and have the capability of causing a serious infection in the host. It should be obvious that the use of the INT technique is not an elixir to answer all the questions you ever had about *Legionella*. It does however provide a technique that allows the gathering of *in situ* information that would otherwise not be available.

III. ECOLOGICAL DISTRIBUTION

As we begin this section, I am reminded of Arthur T. Henrici's comment (1933), "Ecology is essentially a revolt from the analytic and experimental methods of the laboratory, on the ground that these give no adequate conception of the organism as a whole in relation to its environment, and a return to the methods and aims of the old-fashioned field naturalist, freed however from the teleologic bias and with the addition, as far as possible, of the precision of the laboratory". The ecological studies on *Legionella* certainly flowed from the knowledge that the laboratory findings were inadequate to explain the relationship of *Legionella* to its environment. I am sure that I do not agree with all of Henrici's statement but it is certainly a perspective of the microbial ecologist.

If one were to study the location of bacteria in the natural environment it would be noted that most of the bacteria are associated with the interfaces of inert particles and biological surfaces.[60] The existence of life has the requirement of water within a boundary condition. The presence of only water prevents bacterial growth as does the total absence of water. Thus, all bacteria are aquatic and film organisms. So it is with *Legionella*. The bacterium is essentially aquatic not only for the above esoteric reason, but also it is the only habitat from which *Legionella* has been routinely isolated.

Our laboratory has been involved with the question of geographic distribution of *Legionella* since 1978. We have a 35-ft mobile laboratory that affords us the versatility of moving into a variety of environments in order to process samples in the field within minutes of collection so that changes due to storage or shipping are kept at a minumum. Thus physiological ecology can be performed with the minimum of sample alteration. Water samples have been collected from all parts of the country and processed for the presence of *Legionella*. These samples have been primarily from surface waters, including major rivers and large lakes both natural and man-made, both ambient and thermally altered.[4,5] The majority of sampling has occurred in the southeastern U. S. because of our juxtaposition. The story appears to be consistent with aquatic bacteria[61] in that wherever one finds water, one will find *Legionella*. Present techniques allow the researcher to detect *Legionella* at lower concentrations and likewise the habitats available for isolation have increased.[14,16,17,62-74]

Much of the research that has been conducted on the ecological distribution of *Legionella* has been done with the first four serogroups of *L. pneumophila*. Part of the reasoning behind such a study is historical in that *L. pneumophila* was the initial organism isolated. Additionally, these serogroups have composed a significant fraction of the Legionellosis cases in the U.S.[2]

During the past four years, monthly water samples have been taken from 7 fixed stations within a 3000-acre reservoir Par Pond, in South Carolina and processed for the presence, density, distribution, activity and virulence of *Legionella*. Initially this lake was studied because it has been described in the scientific literature for a number of years and a great deal is known about its biological and abiological components.[12,61,75-78] Furthermore, Par Pond has never been associated with an outbreak or a sporadic case of legionellosis.

Par Pond has three parallel arms and samples were routinely collected from the stations shown in Figure 2. The samples were aseptically collected from one meter depth intervals and immediately processed. Figure 3 depicts the thermal regime from which the samples were taken and included samples from temperatures greater than 70°C to ambient temperatures which in winter are at times 4.0°C. Each of the sampling stations was monitored for

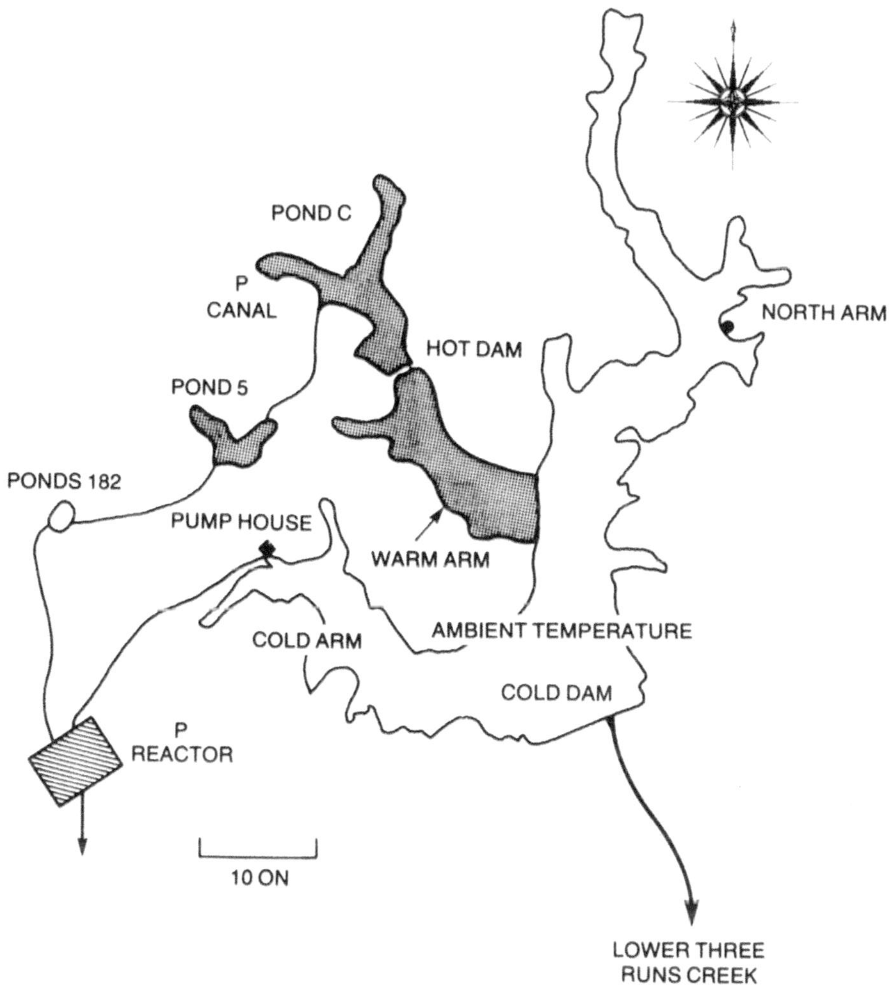

FIGURE 2. Map of the Par Pond System showing overall canals and lake habitats.

physical and chemical properties which included temperature, dissolved oxygen, conductivity, pH, redox potential, secchi disk depth reading (which is a measure of the clarity of the water), total dissolved organic carbon, dissolved inorganic carbon, nitrate, nitrite, total nitrogen, orthophosphate, total phosphate, the amount of algal photosynthetic material present in the sample, and the percentage of respiring *L. pneumophila*.[4,30] The DFA densities were measured and the isolation of *L. pneumophila* from these various habitats was confirmed as previously described.[4,5]

Early on, numerous workers suggested that the culturing conditions for *Legionella* were so restrictive that the organism probably did not exist in unstable environmental habitats. As we know, such is not the case. The point cannot be made enough, the extrapolation of laboratory data into the field and *in situ* conditions can cause great errors! *Legionella* is quite capable of surviving temperatures in excess of 60°C but probably not above 65°C.[30] Recent studies by Yee and Wadowsky[74] reaffirm that laboratory and nonlaboratory studies often provide different answers although the questions are the same.

The initial ecological studies[4,5] demonstrated that *Legionella* was somewhat unusual at least in its temperature relationships, and that nonambient habitats should be considered. The temperature phenomenon was initially studied because of my research and acquaintance with other thermal habitats, particularly those in Yellowstone National Park. Those years

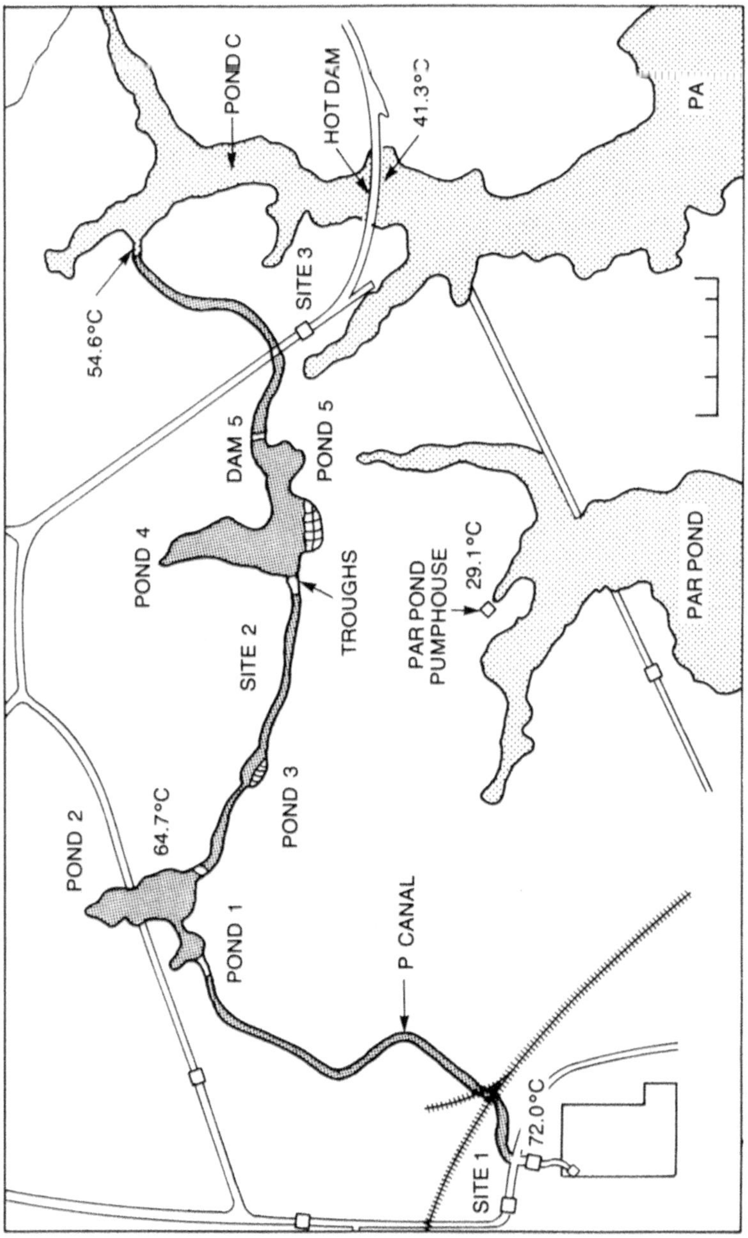

FIGURE 3. Map of the thermal canal portion of the Par Pond System showing *in situ* temperatures and sampling sites.

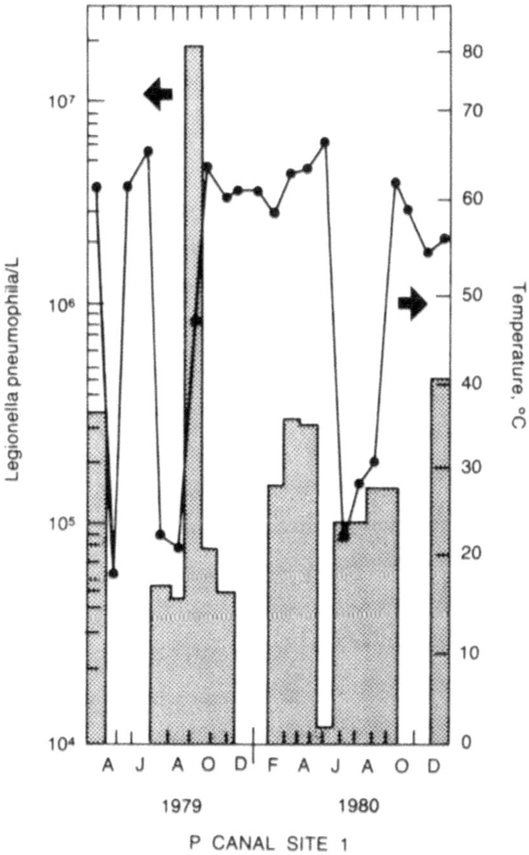

FIGURE 4. Density estimates of *L. pneumophila* serogroups 1 to 4 at site 1 along the thermal canal.

of investigation had led the thermophilic community of scientists to recognize characteristics of bacteria associated with natural thermal environments. One of these characteristics was the general pattern that thermophilic bacteria possessed a large percentage of branched chain fatty acids such as found in *Thermus aquaticus*, *Bacillus caldolyticus* and *B. caldotenax*.[79-81] The studies on the Par Pond System have revolved around the impact of man-made thermal alterations on the microbiological density, diversity, and distribution in these habitats.

The data in Figures 4, 5, and 6 demonstrate the density of *Legionella* with respect to the temperature of the reactor canal habitats during a 22-month period. The data indicate that at the hottest temperatures tested the densities of *L. pneumophila* are depressed. As the temperature of the habitat is lowered by atmospheric cooling, the densities of *L. pneumophila* increase with the greatest densities observed at habitat temperatures between 45 and 55°C. Once the thermal alteration is no longer experienced the seasonal pattern common for the southeastern U.S. predominates (Figure 7). Although temperature plays a role in the distribution of *Legionella*,[4,30] it is not the entire driving force that determines the density or the habitat niche of the bacterium.

One of the desired outcomes of this ecological research is that a type of indicator might serve as a surrogate for *Legionella*. This so far has not been the case. Just as fecal coliforms are not particularly good indicators for fecal pollution,[82,83] it is unlikely that an indicator for *Legionella* will be forthcoming. A number of statistical correlations have been made with the various data sets taken from ecological habitats [134] along with published data from a limited study,[4] however, none of these data have shown any water quality characteristics

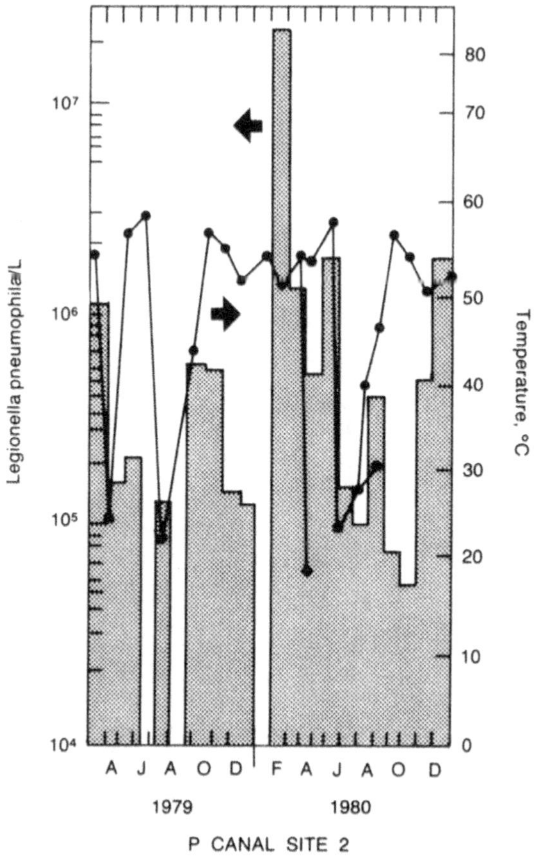

FIGURE 5. Density estimates of *L. pneumophila* serogroups 1 to 4 at site 2 along the thermal canal.

that can be used as a strong predictor of the presence or absence of *Legionella.* Although a wide range of habitats harbor *Legionella,* those aquatic communities that had temperatures of greater than 40°C but less than 60°C had increased frequency of isolation through guinea pigs.

Density estimates of *Legionella* spp. have been made in the Savannah River and its estuary. The data in Table 6 demonstrate the results of a sampling effort along the Savannah River from Augusta, to Savannah, Ga., and out into the Atlantic Ocean. The data indicate that the waters flowing into the coastal area do contain species of *Legionella* but that the densities are reduced once the levels of salinity increase. Samples collected in the open ocean were not even DFA positive for any of the species of *Legionella* tested. The data are preliminary indications that *Legionella* may not be tolerant of marine or estuarine environments.

Ortiz-Roque and Hazen[84] have reported that *Legionella* is part of the freshwater ecosystem in the Caribbean National Forest. The densities of *Legionella* are several orders of magnitude higher than thus far reported. Additionally, *Legionella* has been reported in the estuaries of Puerto Rico and the Caribbean Sea. These data, when confirmed and published, suggest that what is known about *Legionella* in a temperate climate may be entirely different for the organism in the tropical habitats such as Puerto Rico.

Although our studies have centered around man-made thermal habitats, we have sampled some natural warm and hot springs habitats in the U.S. (Table 7). The data indicate that *Legionella* is indeed part and parcel of the habitats that are naturally hot in nature. Many of these natural habitats are alleged to be quite old and certainly they are when compared

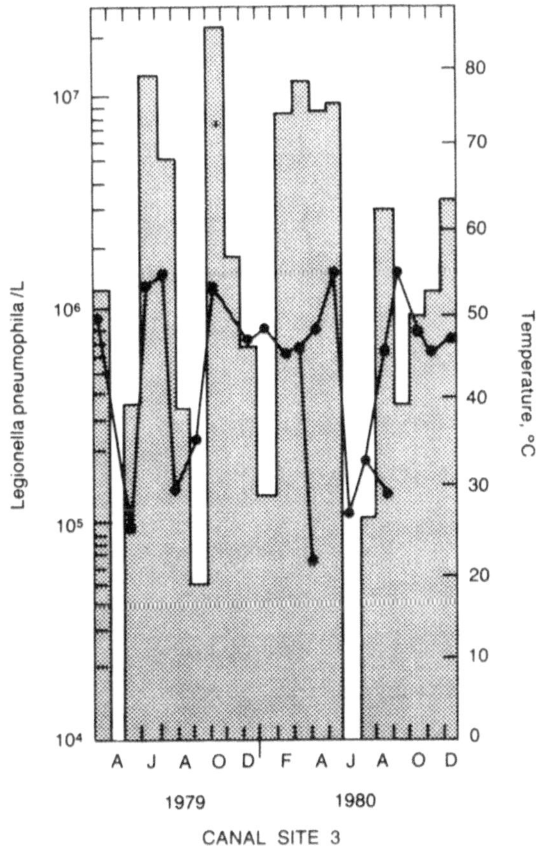

FIGURE 6. Density estimates of *L. pneumophila* serogroups 1 to 4 at site 3 along the thermal canal.

to the man-made habitats of nuclear production reactors. While some of the samples from Yellowstone National Park were collected and processed immediately, others were shipped through the courtesy of Dr. Dave Ward of Montana State University. Generally, the samples contained densities of *Legionella* comparable to those observed in the man-made habitats or in natural habitats throughout the U.S. Some habitats have densities 2 to 3 orders of magnitude above those observed in the natural nonthermal habitats. Isolations from these natural thermal habitats appear more difficult than from man-made thermal habitats on an equal effort sampling basis. The data seem to indicate that the man-made thermal habitats are more likely to contain infectious *Legionella* than the natural thermal habitats. The reason for such findings is unclear but may be related to the cycling of warm water in the man-made aquatic habitats and the stability of natural hot springs.[25]

Tison et al.[32] have shown that the periphyton community in the thermal reactor canals of the Par Pond System changes dramatically in its physiology when the reactor is shut down for refueling or other maintenance. During these thermal removal conditions the algal community is stressed so that the release of extracellular photosynthate is substantially increased. This extracellular organic matter is thus readily available to the bacterial components of the ecosystem such as *Legionella*. The significance of these findings are yet to be fully realized but they may indicate a difference in the virulence of the *Legionella* in these two distinct thermal habitats.

The eruption of Mt. St. Helens has provided lake conditions in which *Legionella* may grow. Initially in 1980 we processed water samples from the various lakes formed or

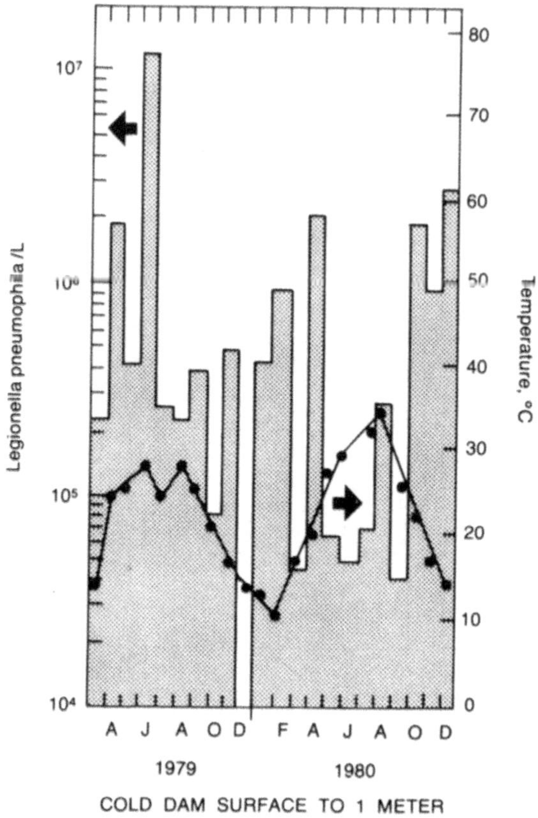

FIGURE 7. Density estimates of *L. pneumophila* serogroups 1 to 4 at site V at an ambient temperature habitat (0 to 1 m) in the Par Pond System.

significantly altered during the eruption. The data in Table 8 demonstrate the densities of *Legionella.* Tison et al. have extended these studies[85] and have shown that these habitats contain densities of *Legionella* that are probably greater than the levels observed in many natural habitats. Likewise the distribution of *Legionella pneumophila* serogroups in Mt. St. Helens habitats are primarily 4 and 6 while other habitats show a preponderance for serogroups 1 and 4.

Terrestrial habitats have remained a potential source of *Legionella* but have not been confirmed as a reservoir for the organism. Epidemiological investigations of the outbreak in 1965 in St. Elizabeth's Hospital in Washington, D. C., have retrospectively implicated the excavation of soil as a possible reservoir and route of infection.[86] During the epidemic a new sprinkler system was being installed in the grounds of the hospital and individuals were infected in a juxtaposition to the implementation of that system. Epidemiologically the excavation sites were implicated but no positive samples for *Legionella* were obtained. It must be kept in mind that 8 days before the epidemic heavy rains and high winds may have played a role in disseminating *Legionella*. This scenario of terrestrial habitats being implicated in the reservoir of *Legionella* has occurred over the past few years and the legend of terrestrial *Legionella* has persisted without evidence.[87,88]

An outbreak associated with legionellosis at the Memorial Union in Bloomington, Ind., had as one of the isolates a positive sample from "soil". These samples were actually mud samples taken from the banks of the small stream called euphemistically the Jordan River. This stream is heavily impacted by discharges from some of the university buildings' steam

Table 6
DISTRIBUTION OF *LEGIONELLA* SPP. IN THE SAVANNAH RIVER ESTUARY 12/15/78 USING DFA

Sample number	Location	Salinity (ppt)	Number/ℓ				Isolation
			L. pneumophila	L. bozemanii	L. micdadei	L. gormanii	
104-4	Augusta, GA Rt #1; River Mile 199.5	0.05	1.1×10^6	1.8×10^4	BD	1.1×10^5	*L. pneumophila*
104-2	Savannah River Plant; River Mile 156.8	0.04	8.9×10^5	1.1×10^4	9.1×10^3	BD	Neg
104-1	Highway # 301 Bridge; River Mile 118.7	0.06	2.8×10^6	3.7×10^5	BD	2.7×10^5	*L. pneumophila*
103-4	Highway × 17 Bridge; River Mile 21.4	0.50	1.4×10^5	BD	BD	2.8×10^4	Neg
103-5	Pilot House; River Mile 0.8	18.0	9.1×10^3	BD	BD	BD	Neg
103-6	River Mouth; River Mile 0.0	18.1	1.1×10^4	BD	BD	BD	Neg
103-7	2 miles from Mouth in Atlantic	34.5	BD	BD	BD	BD	Neg
103-8	3 miles from Mouth of S. River in Atlantic	35.0	BD	BD	BD	BD	Neg

Table 7
DENSITIES OF *L. PNEUMOPHILA* IN SELECTED NATURAL WARM AND HOT SPRINGS IN THE UNITED STATES

Habitat	Location	Temp (°C)	Serogroups number/ℓ				Isolation
			Knoxville	Togus	Bloomington	Los Angeles	
Cyanidium Creek	YNP*	42	BD	BD	BD	BD	NT
Cyanidium Creek	YNP	45	BD	BD	BD	BD	NT
Cyanidium Creek	YNP	50	BD	BD	BD	BD	NT
Cyanidium Creek	YNP	55	2.3×10^8	BD	BD	BD	K
Bath Lake	YNP	36	1.5×10^8	BD	BD	4.1×10^7	K,LA
Octopus Lake	YNP	45	BD	BD	BD	2.7×10^7	Neg
Octopus Lake	YNP	50	BD	BD	BD	BD	Neg
Octopus Lake	YNP	55	6.8×10^7	BD	BD	BD	K
Octopus Lake	YNP	60	9.5×10^7	BD	BD	BD	K
Mineral Hot Springs	Colorado	54	5.8×10^5	BD	BD	4.1×10^4	Neg
Madison River	YNP	13	6.5×10^6	BD	1.1×10^5	9.1×10^3	Neg
Terrace Spring	YNP	48.5	1.1×10^5	BD	BD	6.6×10^8	LA
Old Bath	YNP	36.5	BD	2.8×10^5	1.7×10^4	1.7×10^4	Neg
Obsidian Creek Springs	YNP	35.5	8.4×10^4	7.1×10^4	BD	8.2×10^4	NT
Amphitheatre Springs	YNP	31.3	BD	BD	BD	BD	NT
Tantalus Creek	YNP	28	2.0×10^4	9.1×10^3	BD	9.1×10^3	NT
Gray Lake	YNP	64.5	3.8×10^5	BD	BD	1.1×10^5	Neg
Pool A	YNP	65	BD	BD	BD	BD	NT
Pool A outfall	YNP	45	5.7×10^5	4.0×10^4	BD	8.1×10^4	Neg
White Creek	YNP	51	3.0×10^4	BD	BD	7.1×10^7	Neg
White Creek	YNP	47	BD	BD	BD	6.5×10^5	Neg
White Creek	YNP	41	3.1×10^7	1.1×10^5	4.0×10^4	1.1×10^5	Neg
Lifsey Springs	N. C.	21.3	2.5×10^5	BD	BD	BD	Neg
Thundering Springs	VA	10.3	2.2×10^5	1.6×10^5	BD	1.7×10^4	Neg
Barker Springs	VA	18.5	BD	BD	BD	BD	NT
Warm Springs	GA	31	BD	8.2×10^4	BD	BD	Neg
Parkman Springs	GA	9.5	4.5×10^5	3.8×10^5	BD	8.2×10^4	Neg
Tom Brown Springs	GA	8.5	1.1×10^6	1.9×10^5	BD	1.4×10^5	Neg

YNP = Yellowstone National Park. NT = Not Tested. BD = Below Detection.

Table 8
DENSITY ESTIMATES OF *LEGIONELLA/ℓ* IN WATER SAMPLES COLLECTED SEPTEMBER 1980 FROM LAKES ALTERED BY THE ERUPTION OF MT. ST. HELENS

Sample	Guinea pig isolations	*L. pneumophila* 1-4 (A)	*L. pneumophila* 5-6 (B)	*L. bozemanii* *L. micdadei* *L. gormanii* (C)
Coldwater Lake (55°C)	A	9.1×10^6	5.8×10^7	5.4×10^7
Coldwater Lake (78°C)	Neg	BD	BD	BD
North Cold Water	Neg	BD	BD	8.4×10^7
Castle Lake	A,B	1.9×10^7	2.2×10^6	1.3×10^8
Ryan Lake	Neg	9.1×10^5	9.6×10^6	1.2×10^7
Spirit Lake	A	6.8×10^6	2.2×10^7	5.7×10^7
North Fork Tuttle River	Neg	5.0×10^6	9.6×10^6	1.5×10^8

lines.[89] Thus the presence of *Legionella* is not surprising. The point is that the samples were heavily influenced by the aquatic community and it is not possible to use these samples as evidence for terrestrial sources of *Legionella.*

Numerous researches have attempted to isolate *Legionella* from the terrestrial habitat without success.[135] It is indeed true that within soil organisms, such as earthworms, arthropods, and the like, a large number of *Legionella*-like bacteria can be observed by the direct fluorescent antibody technique but the organisms have not been isolated and the necessary confirmation is lacking.[50] Thus one is left with the hypothesis that *Legionella* may well be a member of the terrestrial community but it is in fact not culturable from this "source". The data remain that *Legionella* is a part of the aquatic community and has thus far not proven itself to be part of terrestrial habitats

The American Society of Heating, Refrigeration and Air-Conditioning Engineers, Inc. (ASHRAE) have determined that certain natural and man-made habitats are amplifiers of *Legionella* and thus deserve special consideration.[90] An amplifier is defined. . ."as any natural or man-made system which accepts an input of viable *Legionella* from some source and provides favorable conditions for the accumulation or growth of *Legionella*". Such amplifiers may well include but are not limited to cooling towers, humidifiers, hot water systems, lakes, aquatic plants, algae, etc. The significance of these amplifiers is determined by the source of their water for input and the dissemination of the *Legionella* that is amplified. If the source were high in *Legionella* then the amplifier would be readily seeded and inoculated. If the dissemination of the *Legionella* were to expose a large population, then the amplifier is potentially more hazardous than one that is isolated from the populous. These two facts are or can be embodied in one such amplifier, the cooling tower.

Since the primary route of infection is via aerosols,[87] it is unusual that so little has been done to detect, quantify and determine the distribution of *Legionella* in ambient air samples. The juxtaposition of towers can be most important, since meteorological conditions can play a large role in the dissemination of the bacterium. The closer the tower is to the populous area, the more likely and frequently will individuals become exposed to the dissemination from the towers. The question is whether that exposure has any consequences. Meterological considerations using simple modeling techniques without any verification with respect to *Legionella* indicate that densities of the organism can be detected downwind for a considerable distance.[91] Every cooling tower that I have sampled or have been associated with has had detectable levels of *Legionella*[92,93] and such observations are supported by other workers.[6,94]

Table 9
POTENTIAL OR PROVEN MAN-MADE SOURCES OF *LEGIONELLA*

Evaporative condensers	Boilers
Taps—hot and cold water	Fountains
Humidifiers	Cooling towers
Nebulizers	Aquariums[a]
Hot water tanks	Saunas[a]
Potable water	Dehumidifiers[a]
Shower water and shower heads	Respirators[a]
Whirlpools	Swamp coolers[a]

[a] Potential sources.

Data from a number of sources indicate that a threshold concentration may be required to cause infection.[5,30,95-98] Although cooling towers may harbor *Legionella,* the mere presence of the organism is not cause for an outbreak. There well may be a density level that *Legionella* may have to achieve in a cooling tower before the system is considered hazardous. An arbitrary density of 10^8 *Legionella* per liter has been chosen as an upper limit. [92] This arbitrary level was established based on circumstantial lines of evidence, and thus may need revision as more data are available. It is an area of much needed research. The point must be made that the mere presence of *Legionella* in an amplifier has very little relevance as to the status of that amplifier with respect to the final dissemination of the bacterium.

The summary Table 9 for this section indicates the various man-made habitats that are proven or are potential habitats for *Legionella.* Such habitats are places where *Legionella* have been isolated or are potential habitats. Although the Environmental Protection Agency has an overarching principle that water for human use is to be free of pathogens, consumers need to be aware of potential niches of *Legionella.* Thus individuals at high risk to infection and thus susceptible to legionellosis will not be unduly exposed.

IV. ENVIRONMENTAL ASSOCIATIONS

Based on the development and the utilization of the various techniques that have previously been described, the stage was set to investigate the interactions of *Legionella* with its habitat and the other micro- and macroorganisms that are present in these habitats. These associations are not unique to *Legionella* but are indeed part of the functioning of aquatic ecosystems. As surely as night follows day, there is a cycling and a seasonality to the ecological world. The nutrients that are made through the process of photosynthesis are utilized by the chlorophyll-containing organisms and the products that are formed are available for the nutrients of others either directly or through the process of decay and death which is the story of all creation. As the organisms experience these times of growth, nutrient utilization, dissolved organic matter release, and finally death they provide nutrients and growth factors for the rest of the aquatic community. *Legionella* is part of this community.

I first became aware of the possibility that *Legionella* may be associated with the aquatic community because of the epidemiological data that appeared in *Morbidity and Mortality Weekly Report.*[99] The data demonstrated that the cases of legionellosis followed a seasonal trend with the highest incidences occurring in the late summer to early fall. Such periodicity follows the trend of many aquatic bacteria that respond to the various factors in the aquatic habitat, i.e., temperature, solar radiation, photosynthesis, etc.

Microscopic observation of *Legionella* in the natural habitats invariably demonstrated the

Table 10
BIOVOLUMES (MEANS ± 1 SD) OF MAJOR[a] ALGAL SPECIES AT SITE 3 ON FIVE SAMPLING DATES

	Biovolume (mm^3/cm^2)				
	2/12[b]	**3/13[c]**	**3/22[d]**	**4/06[e]**	**4/28[f]**
Oedogonium sp. #1	17.6 ± 1.0	4.0 ± 1.8	5.7 ± 0.1	6.7 ± 0.1	0
Oedogonium sp. #2	4.7 ± 0.6	0.6 ± 0.2	0.1 ± 0.2	0.9 ± 0.1	0
Navicula confervacea	<0.1	2.3 ± 1.0	0	<0.1	0
Synedra delicatissuma	0	0.8 ± 0.2	0.1 ±<0.1	0	0
Phormidium sp.	0.1 ±<0.1	0.1 ±<0.1	0.7 ± 0.1	0.5 ± 0.1	1.0 ± 0.1
Fischerella sp.	<0.1	0.1 ±<0.1	0.1 ± <0.1	0.1 ±<0.1	0.1 ±<0.1
Oscillatoria sp.	0	0.1 ±<0.1	0.1 ±<0.1	0.8 ±<0.1	0.8 ± 0.2
Total algae	22.7 ± 0.7	11.1 ± 4.5	7.1 ±< 0.1	9.3 ± 0.6	1.9 ±<0.1

[a] Comprising at least 5% of the total algal biovolume in one or more samples.
[b] After one month exposure to thermal conditions. Temperature on sampling date = 40°C.
[c] After eight days exposure to ambient temperatures. Temperature on sampling date = 19°C.
[d] After two days exposure to resumed thermal conditions. Temperature on sampling date = 40°C.
[e] After two weeks exposure to thermal conditions. Temperature on sampling date = 44°C.
[f] After five weeks exposure to thermal conditions. Temperature on sampling date = 45°C.

presence of chlorophyll containing algae. Typical color photomicrographs have been published.[5] This association is one of the stories that I believe *Legionella* would like to tell us if we have ears to hear, because within the aquatic community a variety of associations occur that we probably know virtually nothing about. Table 10 demonstrates the species composition of the algae associated with *Legionella* at 45°C in the thermal canal.[100] Similar investigations have not been conducted extensively in cooling towers to define the types of algae associated with *Legionella*.[8]

The second point that the algal *Legionella* interaction makes is the fact that the sampling of *Legionella* in the natural habitat can be very patchy leading to meaningless data if the samples are not collected to minimize patchiness. This means that grab samples are not appropriate for detailed study. Generally for the *Legionella* density estimates using the techniques of the DFA, the reproducibility with 4 ℓ water samples is good with a standard error of 5 to 10%. With triplicate water samples the lower limits of detectability are 9 *Legionella* per milliliter of sample. Where practical triplicate samples are desired, duplicate samples are required. Table 11 shows the reproducibility of the FA technique with respect to the density estimates of *Legionella*. Most studies have concentrated on the use of the first four serogroups of *Legionella pneumophila* since these are the *Legionella* that have been those most associated thus far with the various outbreaks across the world.[4]

The use of the DFA technique is best served if the samples are treated with a 2% solution of gelatin[4,41,45] to lower the background, and if individual serogroups are used on the environmental samples. Within the various man-made habitats a variety of chemicals exist which will tend to increase the background fluorescence at times, particularly if that FA is pooled sera. The reason for such is unclear, it simply happens. Part of the problem is related to the generally higher background fluorescence associated with the ecological samples when compared to the relatively clean samples of the clinical specimens. This high background is generally overcome through the use of a gelatin pretreatment followed by a final rinse in a 5% sodium pyrophosphate solution.

Through laboratory experiments there has been some data that seem to indicate that the virulence of *Legionella* is affected by its associations with the algal component.[101] The data shown in Table 12 suggest that *Legionella* has an increase in virulence when grown in

Table 11
REPRODUCIBILITY OF THE DFA TECHNIQUE FOR THE ESTIMATE OF *L. PNEUMOPHILA* IN TRIPLICATE WATER SAMPLES COLLECTED FROM THREE DIFFERENT COOLING TOWERS

	Number of *L. pneumophila*/ℓ				
Cooling Tower	**Knoxville**	**Togus**	**Bloomington**	**Los Angeles**	**Total**
A_1	1.09×10^5	6.99×10^6	BD[a]	BD	7.10×10^6
A_2	1.64×10^5	6.33×10^6	BD	BD	6.49×10^6
A_3	1.37×10^5	6.25×10^6	BD	BD	6.39×10^6
B_1	5.19×10^5	1.86×10^6	5.46×10^4	BD	2.43×10^6
B_2	5.46×10^5	2.01×10^6	5.97×10^4	BD	2.62×10^6
B_3	4.10×10^5	1.88×10^6	1.09×10^5	BD	2.40×10^6
C_1	8.72×10^6	3.71×10^6	1.05×10^6	5.11×10^7	6.46×10^7
C_2	8.10×10^6	4.07×10^6	1.21×10^6	4.89×10^7	6.23×10^7
C_3	8.52×10^6	4.81×10^6	1.13×10^6	4.90×10^7	6.35×10^7

[a] BD = Below Detectable Limits of 9.10×10^3/ℓ.

Table 12
RESPONSE OF GUINEA PIGS TO I.P. INJECTIONS OF *L. PNEUMOPHILA* (SRP-22) UNDER CO-CULTURE CONDITIONS WITH *FISHERELLA* SP.

Culture	Number of *L. pneumophila* injected I.P.	Number of *guinea pigs* infected[a]/injected
Co-culture with *Fisherella* sp.	10^5	3/5
	10^6	5/5
	10^7	5/5
	10^8	5/5
BCYE grown *Legionella*	10^5	0/5
	10^6	0/5
	10^7	3/5
	10^8	5/5

[a] Infected guinea pigs were determined by DFA examination of necropsied spleen, liver, and interperitoneal fluid.

association with the algae as compared to the cultures grown on buffered CYE agar alone. These experiments have not been repeated. The problem in part is that it is very difficult to separate *Legionella* completely from the algae and their products. This lack of complete separation means that some of the algae and possibly toxic products may be transferred into the guinea pigs. Such products may help dispose the guinea pig host rather than the increased virulence of *Legionella*. The hypothesis is intriguing in that if the virulence of *Legionella* is in fact changed in the presence of certain algae, then the association between *Legionella* and algae in a cooling tower is far more significant.

The first observations of the association with the thermophilic blue-green algae, *Fisherella*, was followed up by a graduate student, David Tison. Dr. Tison was able to deomonstrate quite nicely that *Legionella* could receive all of its nutrients and requirements for growth through this association, and once that association was broken the ability of *Legionella* to express its wide range of physiology was lost.[101] In association with the high temperature

Table 13
UTILIZATION OF *FISCHERELLA* SP. ^{14}C-LABELED DOC BY *LEGIONELLA PNEUMOPHILA* (TOGUS 1)

RESULTS ARE ± 1 S.D. FROM THE MEAN OF TRIPLICATE SAMPLES

Incubation time (hr)	Incorporated	Mineralized	% Min
20	7,144 ± 714	6,428 ± 206	37 ± 5
43	13,131 ± 3,061	11,284 ± 2,046	35 ± 9
51	10,655 ± 1,431	47,719 ± 6,310	63 ± 12

Fisherella sp., *L. pneumophila* was readily grown at temperatures up to 45°C with generation times of 2.7 hr. This growth continued until the dissolved organic carbon that was supplied by the algae was exhausted. Experiments were conducted in minimal salts medium so that all the organic matter utilized by *Legionella* was derived from the algae themselves. This was nicely demonstrated by the incubation of the co-cultures in the dark with no measureable growth of *Legionella* as well as in the presence of DCMU (3-(3,4 dichlorophenyl) l-dimethyl urea) which inhibits photosystem II of the algae.[26]

The latter part of Dr. Tison's thesis was involved with the demonstration that *Legionella* could use radioactively derived photosynthate products from *Fisherella*. Table 13 is taken from Tison's thesis and demonstrates that a significant amount of the radioactively labeled compounds are actually incorporated and utilized as energy sources for the *Legionella*. How far reaching these results are is still unclear but the data suggest that the presence of algae or their products will support the growth and survival of *Legionella*. This has implications for the users of finished water systems in that although the algal component of the ecosystem may not be present, the products of the algae may well be there and provide the necessary nutrients for growth of *Legionella*. This may be part of the explanation why *Legionella* is found in places where algae are not, such as tap water. One of the hypothesis for the ubiquity of *Legionella* is their ability to obtain their nutritional requirements from a wide variety of aquatic community.[102]

Other researchers have demonstrated that an association does occur between a particular kind of algae (blue-green) and a variety of heterotrophic bacteria in the aquatic communities.[103-106] These studies have been supported in part by direct observation but more elegantly by the use of radioactive techniques that have demonstrated the transfer of radioactive material produced during photosynthesis by the algal component to the bacterial assemblages.[106,107] The organic material that is transferred or extracellularly produced by the algae is known as DOM. These organic substances contain a variety of compounds including amino acids, organic acids, polysaccharides, carbohydrates, enzymes, vitamins, and lipids.[108,109]

Kuentzel[110] has suggested that there is commensoral relationship between bacteria and algae in the natural habitat in that the bacteria provide the necessary CO_2 that is often a limiting nutrient, particularly during times of surface algal blooms. The CO_2 is derived from bacterial respiration.[102,104,111] Additionally, the algae provide to the bacterial components an array of extracellular carbon and nitrogen compounds as well as the necessary oxygen to perform the metabolic processes. Several species of opportunistic pathogenic bacteria and indicator species of fecal pollution have been demonstrated to utilize the extracellular components of algae for growth.[112]

Subsequent studies by Ellen Braun at RPI have been less fruitful simply because the problem was far more difficult.[8] The studies incorporated the ground work laid by Tison but attempted to delineate the nature of the substances utilized by the *Legionella* that were derived from the algae. Part of the lack of success may have been the use of lower temperature blue-green algae, *Anabaena spiroides,* rather than *Fisherella.* Dr. Braun's studies indicated that a small molecular weight compound, <500 mol. wt., was responsible for the majority of the uptake observed in the pure cultures of *Legionella* supplemented with the radioactive algal DOM. Attempts to purify and characterize the substance were not successful, although the demonstration that ^{14}C-labeled photosynthate was incorporated into the macromolecules of *L. pneumophila* along with the concomitant utilization and respiration to $^{14}CO_2$ was clear.

Although Pearl has suggested that organic nitrogen compounds were leaked from the heterocystic containing blue-green algae,[115] data obtained by Braun indicate that no difference as observed in the growth rates of *Legionella* both in the presence or absence of heterocystic, *Anabaena spiroides.* It appears that it is very difficult to grow *Legionella* on extracellular products excreted by the various algae in the absence of the algae themselves. A close juxtaposition between the algae and *Legionella* is necessary for growth. Using dialysis techniques Braun was able to show that the active compound(s) has a molecular weight of less than 1000 daltons and is relatively insensitive to a boiling water bath for 60 min, but cannot be autoclaved.[8] It is appropriate to end this section of the chapter with some consternation. The findings of the association between *Legionella* and the various forms of algae are interesting and illuminating in that they suggest the various ways in which *Legionella* may survive the environmental parameters in which it finds itself. The initial thinking was that this observed association would not only tell us about the distribution of *Legionella* but also illuminate one as to the compounds it favored in the natural habitat. Knowing these compounds it was thought that specific cidal or static approaches could be forthcoming. To date this area of the research has not produced the desired fruit, although the fruit has been forthcoming in other ways.

V. AQUATIC MACROPHYTES

Although algae are cosmopolitan in their distribution, so are aquatic plants and particularly submerged aquatic macrophytes. Within the Par Pond System, which has been studied extensively, a wealth of knowledge is known about the submerged macrophytes. In the observations of the *Legionella* distribution in this lake it has become apparent that the major zones of concentration are in the surface waters and often in juxtaposition to aquatic plants or in association with the algae as has already been discussed. Macrophytes function in the shallow area of the lake (littoral zones) by providing cover for the macroscopic community and substrates for attachment to the microscopic community.[113] Additionally, the macrophytes are important in the cycling of nutrients in that they have the ability to extract nutrients from the underlying sediment with the subsequent release of these nutrients to the surrounding water either via excretion or decomposition. This latter function of excretion and decomposition of submerged macrophytes has intrigued us in the same way as the algae. Previous work in relation to the bacterium *Aeromonas hydrophila* has demonstrated that when the aquatic macrophytes die back each fall, the numbers of *A. hydrophila* subsequently increase as much as 3 to 4 logs close to the decaying plants, while in the open waters 10 m away the numbers of the organisms are much reduced.[12,31]

Within the Par Pond System the submerged macrophytic community is dominated by the presence of three species of aquatic plants, *Myriophyllum spicatum, Najas quadalupensis,* and *Eleocharis acicularis.*[114] Since these plants are present all over the U. S.,[115] any observed association between these plants and *Legionella* may help to explain the widespread distribution of *Legionella. M. spicatum* were collected from Par Pond as well as lakes in New York, Vermont, North Carolina, and Virginia. The plants (100 g) were homogenized in a Waring blender in 1 ℓ of filter sterile habitat water and the presence or absence of *L.*

Table 14
PRESENCE OF *L. PNEUMOPHILA* IN ASSOCIATION WITH THE SUBMERGED AQUATIC MACROPHYTES

Macrophyte	State	*L. pneumophila*/ℓ	Isolations
M. spicatum	S.C.	5.0×10^7	+
M. spicatum	S.C.	3.8×10^7	+
N. quadalupensis	S.C.	BD	−
N. quadalupensis	S.C.	BD	−
E. acicularis	S.C.	BD	−
E. acicularis	S.C.	BD	−
M. spicatum	N.Y.	3.8×10^7	+
M. spicatum	N.Y.	8.3×10^7	−
M. spicatum	N. C.	3.0×10^5	−
M. spicatum	N.C.	8.5×10^5	−
M. spicatum	Va.	1.7×10^7	−
M. spicatum	Va.	7.0×10^7	+
Water	S.C.	3.0×10^5	NT
Water	N.Y.	4.7×10^6	NT
Water	N.C.	5.5×10^4	NT
Water	Va.	3.1×10^5	NT

NT = Not Tested.
BD = Below Detection.

pneumophila was determined by DFA and then followed by guinea pig injections and isolation onto BCYE medium as previously described. The samples were all taken during the latter part of September when the decaying process in the plants was very evident. Table 14 indicates that the blended samples contained significant amounts of *L. pneumophila*. The data suggest that certain submerged aquatic macrophytes provide nutrients which enhance the densities of *Legionella*.

Braun has observed a similar phenomenon in Lake Champlain near Burlington, Vermont.[8] It is interesting to speculate since the outbreaks that have occurred in the Burlington area have been linked to amplifiers that take their source water from the city water system which derives its water from Lake Champlain. From other reports[15,16,33,62,68-72] it is clear that *Legionella* survives the present water treatment process for many major cities and Burlington appears to be another such city. If the time were taken to evaluate the water systems of cities across the nation, the story would be essentially the same. *Legionella* survives water treatment processes for finished water systems.

There are indeed other associations that have not been investigated that need to be mentioned. Tyndall and associates[116] have demonstrated that virulent *Legionella* can have an inhibitory effect on the amoeba, *Naegleria*. *Naegleria* has been shown to be associated with thermal effluents and has been isolated and characterized from the same habitats and some of the same sampling stations as *Legionella*.[117] Tyndall has also demonstrated that this amoeba is part of the microbial population in cooling towers.[116] Such findings have led Robotham to postulate that the difference between the pneumonic form of legionellosis and the flulike symptoms is possibly due to the presence of *Legionella* inside a host such as an amoeba.[118]

Tyndall[134] has shown that *Naegleria lovaniensis* and *Acanthamoeba royreba* can both utilize *L. pneumophila* as a sole nutrient source. Some of the amoeba cultures became chronically infected with *Legionella* to the extent that *Legionella* were released from the amoeba. Neither the virulence of the amoeba nor the virulence of the *Legionella* were altered by the association in the co-culture situation with respect to guinea pig infection. This association between two pathogenic organisms has not been clearly studied, but should offer both an interesting as well as a productive study.

Collins et al.[119] have examined over 2800 sera from both humans and animals for the presence of antibodies against the four serogroups of *Legionella pneumophila*. It is interesting that the survey included both domestic as well as wild animals. The sera from horses demonstrated the highest level of antibodies of the sera tested with 31.4% containing titers greater than or equal to 1:64. It is also interesting to note that the horses with greater age also had higher prevalence of *Legionella* antibody. Secondly, the data indicate that the horse population from Colorado is generally lower in its percentage of antibodies against *Legionella* than the population from Pennsylvania. On the other hand, wild animal populations remained unexposed. The data are interesting and may suggest that certain practices with the domesticated animal population may cause the population to be exposed to *Legionella*. The data suggest that the activities of man may indeed produce habitats that are conducive to the growth of a bacterium which has been part of our aquatic habitat for a long time, but has only recently found a point of dissemination with the increasing use of amplifiers.

A final area of association that has been investigated is the area of water fowl that inhabit the Par Pond System. The migratory patterns of the North American Coot are not part of this discussion but it is sufficient to say that these water fowl invade the Par Pond habitats during the late fall and leave the water system during the spring of the year. During that time they feed on the submerged aquatic macrophytes within the system as previously described. Studies were conducted to determine whether the coots coming into the system were potentially carrying *Legionella* and whether the activity of eating the *Myriophylum* would cause an increase in the density of *Legionella* inside the tissues or an increase in the titer of antibodies against the first four serogroups of *L. pneumophila*. The study based on only 40 coots, 20 taken before the feeding and 20 taken after the 5 months on the lake, showed no differences. We were unable to see any DFA-positive cells in any of the necropsied tissues nor were we able to detect the presence of any antibodies to *Legionella*.

VI. SURVIVAL AND DISINFECTION STUDIES

During this chapter I have tried to demonstrate and document the fact that *Legionella* is a ubiquitous aquatic organism that is able to survive and maintain itself in a variety of habitats under a multitude of environmental parameters. The point of this next section is to describe some of the studies that have been conducted in order to try to assess its survival and control the levels of *Legionella*.

Through the years ecological research has led to the use of a variety of techniques to ask and to answer a variety of questions. One of the techiniques that has been used with some success is the diffusion membrane chamber.[10,31,82,83] The chambers allows one to sample repeatedly a known population of organisms over a given period of time and evaluate the response of those organisms with respect to the habitat in which the chambers have been placed. These chambers were initially described by McFeters[120] and have been modified for both lakes and deep water studies[121] as well as for streams and rivers. The chambers retain a bacterial population for sequential sampling while exposing the organism to *in situ* conditions.

Diffusion membrane chambers have been used in the Par Pond System for *Legionella*.[122] These studies have included placing the first four serogroups of *L. pneumophila* in chambers, fitted with 0.2 nm Nucleopore filters, at various temperatures in the natural habitats and observing the change in *Legionella* density over time. Under these parameters the data in Figures 8 and 9 show that there is a differential in survivability of the various serogroups of *L. pneumophila*. At the higher temperatures *L. pneumophila* serogroup 1 survives better than any of the other serogroups tested. At ambient temperatures the survival of all the serogroups are less. The data indicated that all the serogroups of *L. pneumophila* may not

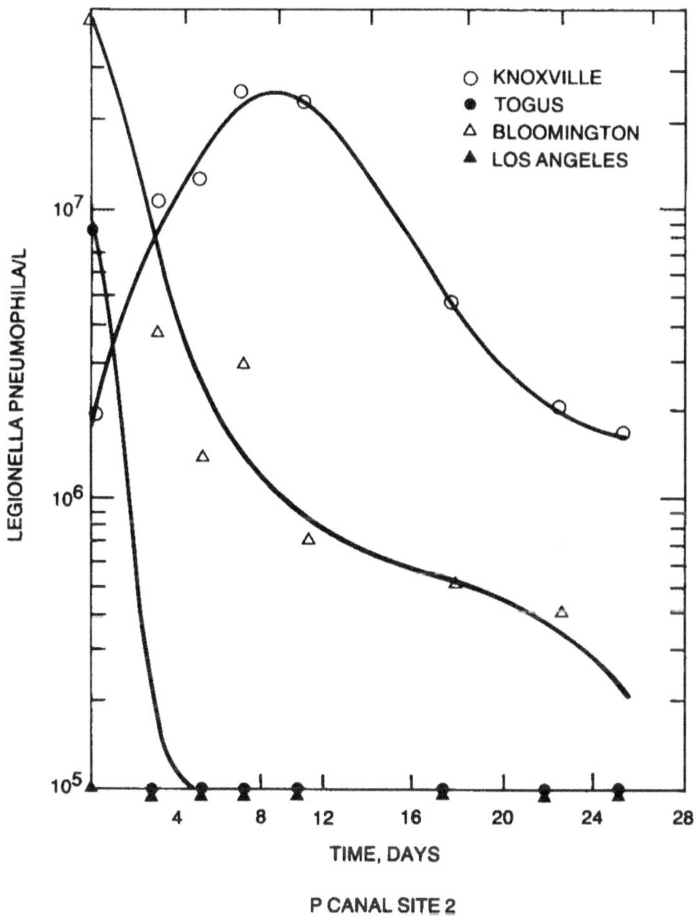

FIGURE 8. Survival of *L. pneumophila* serogroups 1 to 4 in diffusion chambers *in situ* at 55°C.

act the same in the natural environment and that some of the serogroups may be selected for under appropriate environmental conditions.

Subsequent studies were undertaken to determine whether *L. pneumophila* isolated from the 55°C habitats were able to grow at that temperature provided the nutrients for growth were present.[122] Experiments were conducted whereby the *L. pneumophila* were placed in chambers with *Myriophyllum spicatum* and placed at 55°C. Samples were taken over several weeks and the densities of *Legionella* measured by DFA. The data in Figure 10 indicate that *Legionella* in the chambers containing the proper nutrient sources, *M. spicatum*, were able to survive and grow at the indicated temperatures but the *Legionella* not in close juxtaposition with the aquatic macrophyte was not able to survive as well at these temperatures once indigenous nutrients were exhausted.

As pointed out earlier it is not only the warm water habitats that have been associated with *Legionella* but it is the ambient waters as well. *Legionella* has been isolated from shopping mall aesthetic fountains, sewage treatment plants, domestic finished water at the cold water tap, ambient temperature air wash systems humidifiers and de-humidifiers alike, and the list will go on. The point is that the bacterium is aquatic in nature and thus any place that water exists the organism may well be there also.

This then places the burden somewhat on the user in that the user of the water must be aware of the fact that *Legionella* will survive the various water treatment schemes that are

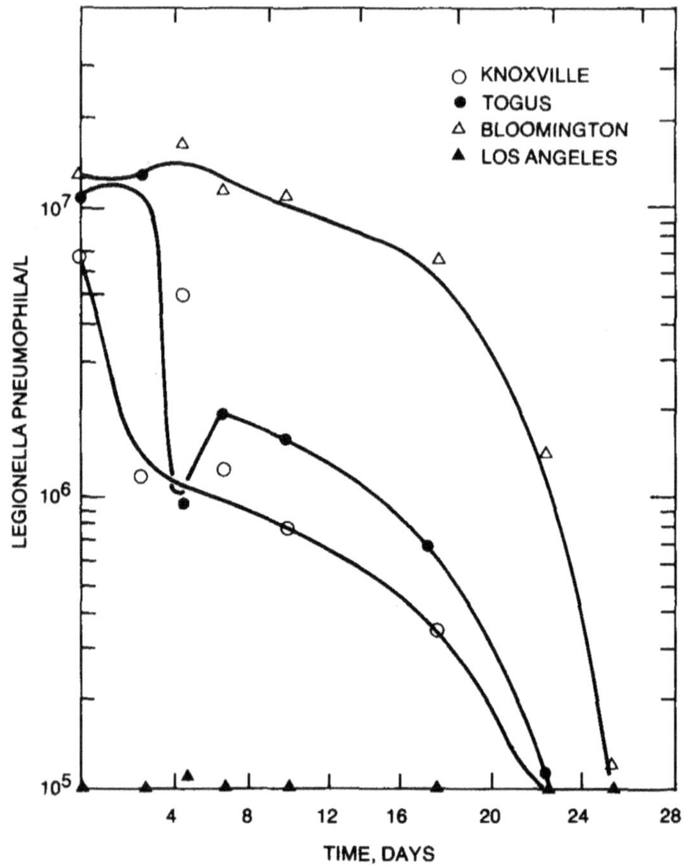

FIGURE 9. Survival of *L. pneumophila* serogroups 1 to 4 in diffusion chambers *in situ* at 23°C.

now employed so that the organism may readily enter into the various systems such as cooling towers, humidifiers, and the like that are considered amplifiers by the ASHRAE. One of the overarching principles of the EPA is the fact that the water that is used by humans for consumption will be free of pathogenic organisms. Such a declaration is noble but one wonders at the practicality of it particularly in the light of the various habitats selected by *Legionella*.

Just as *Legionella* occupies selected niches and habitats, so does man. One of the questions pursued in this ecological research is to determine whether the niches of *Legionella* and the niches for man overlap. If they do, and they do, then the next question has to be the assessment of the significance of that overlapment. Does the mere presence of *Legionella* in the combined niche cause an undue risk and thus needs to be avoided? That is a question for which no data really exists. The thinking suggests that common niches have existed for very long periods of time and only recently have we identified a problem or has a problem become apparent. The mere presence of *Legionella* in a habitat makes no statement as to the quality of that habitat or to the restrictions to be placed on that habitat. It does, however, indicate that one needs to be aware of the potential presence and use the habitat in a way that will not unduly propagate the organism nor cause its dissemination to a susceptible host.

Numerous studies conducted in the laboratory have indicated that a number of biocidal agents are both static and cidal against *Legionella*.[123-126] In the laboratory low concentrations of quaternary ammonium compounds, phenolics, glutaraldehyde, formaldehyde, and hy-

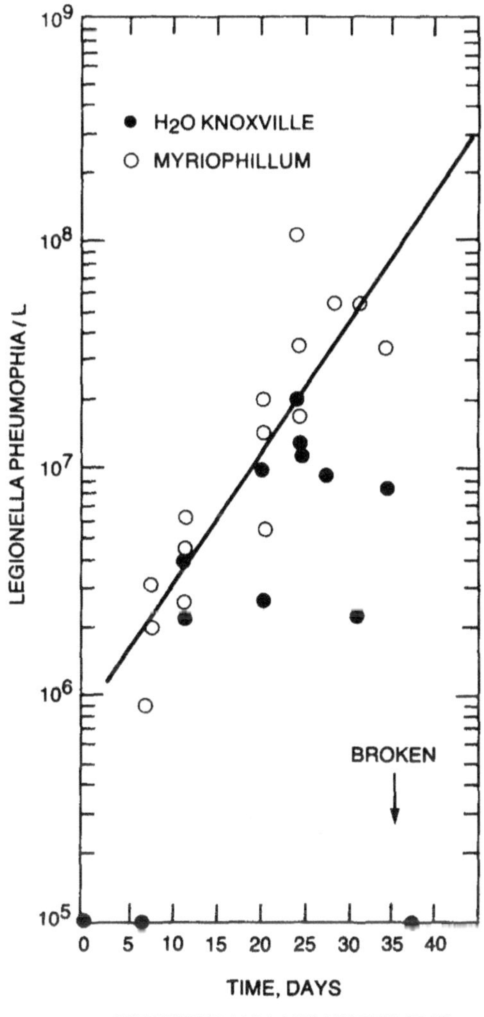

FIGURE 10. Growth of *L. pneumophila* serogroup 1 in diffusion chambers *in situ* at 58°. Chambers contained the macrophyte, *Myriophillum spicatum* (o—o) and source sterile source water (•—•).

pochlorite were effective against *L. pneumophila*.[126] Skaliy et al.[124] demonstrated that biocides containing hypochlorite, 2,2 dibromonitrilo-propionamide, or a combination of quaternary ammonium salts and isopropanol were cidal for *L. pneumophila* suspended in tap water, while isothiazolone, thiocarbamates and chlorophenols were less effective. Grace et al.[123] showed that biocides containing a combination of quaternary ammonium salts and bis (tri-*n*-butyltin) oxide was effective in killing *L. pneumophila* serogroup 1 at 1/25 of the recommended dosage.

A comprehensive study was conducted by Reg Soracco in my laboratory involved the use of 12 commercially available biocides and 9 different strains of *Legionella pneumophila*.[125] The results of the testing are described in Table 15 taken from Dr. Soracco's thesis. The data indicate that the biocides containing tributyltin oxide and quaternary ammonium compounds were the most effective in controlling *Legionella* in the laboratory. Zedler et al.[127] found that organotin compounds are synergistic with the quaternary salts against both Gram-negative and Gram-positive bacteria. The point should be obvious from the above discussion. All these results were obtained in the laboratory and have no value for field situations.

Table 15
EFFECTIVENESS OF THREE CONCENTRATIONS OF EACH BIOCIDE ON GROWTH AND VIABILITY OF THE SELECTED *LEGIONELLA PNEUMOPHILA* ISOLATES

		L. pneumophila isolate																	
		SRP 3		SRP 6		SRP 8		SRP 10		SRP 22		K 1		T 1		B 2		LA 1	
Biocide	Concentrations	E	S,C	E	S,C	E	S,C	E	S,C	E	S,C	E	S,C	E	S,C	E	S,C	E	S,C
	X	+	C	+	C	+	C	+	C	+	C	+	C	+	C	+	C	+	C
H-130	0.1X	+	C	+	C	+	C	+	C	+	C	+	C	+	C	+	S	+	C
	0.01X	−	NT	−	NT	−	NT	−	NT	−	NT	+/-	S	−	NT	+	S	+	S
	X	+	C	+	C	+	C	+	C	+	C	+	C	+	C	+	C	+	C
AG420	0.1X	+	C	+	C	+	C	+	C	+	C	+	C	+	C	+	C	+	C
	0.01X	+	C	+	C	+	S	+	S	+	C	+	C	+	C	+	C	+	C
	X	+	C	+	C	+	C	+	C	+	C	+	C	+	C	+	C	+	C
AG480-F	0.1S	+	S	+	C	+	C	C-	NT	+/−	C	−	NT	+	S	+	C	−	NT
	0.01X	−	NT	+	NT	−	NT	C-	NT	−	NT	−	NT	−	NT	+	S	−	NT
	X	+	C	+	C	+	C	+	C	+	C	+	C	+	C	+	C	+	C
CSA	0.1X	+	S	+	S	+	C	+	C	+	C	+	C	+	S	+	S	+	S
	0.01X	+	S	+	S	+	C	+	C	+	C	+	S	+/−	S	+	S	−	NT
	X	−	NT	−	NT	−	NT	−	NT	−	NT	−	NT	−	NT	−	NT	−	NT
DCF	0.1X	−	NT	−	NT	−	NT	−	NT	−	NT	−	NT	−	NT	−	NT	−	NT
	0.01X	−	NT	−	NT	−	NT	−	NT	−	NT	−	NT	−	NT	−	NT	−	NT
	X	−	NT	−	NT	−	NT	−	NT	−	NT	−	NT	−	NT	−	NT	−	NT
10D	0.1X	−	NT	−	NT	−	NT	−	NT	−	NT	−	NT	−	NT	−	NT	−	NT
	0.01X	−	NT	−	NT	−	NT	−	NT	−	NT	−	NT	−	NT	−	NT	−	NT
	X	+	C	+	C	+	C	+	C	+	C	+	C	+	C	+	C	+	S
KDC	0.1X	+	C	+	C	+	C	+	C	+	C	+	C	+	C	+	C	+	S
	0.01X	+	S	+	C	+	S	+	C	+	C	+	S	+	S	+	S	+/−	NT
	X	+	C	+	C	+	C	+	C	+	C	+	C	+	C	+	C	+	C
KFN	0.1X	+	C	+	C	+	C	+	C	+	C	+	C	+	C	+	C	+	C
	0.01X	−	NT	−	NT	−	NT	−	NT	−	NT	−	NT	−	NT	+	S	−	NT
	X	+	S	−	NT	+/−	S	−	NT	−	NT	+/−	S	+/−	NT	+	NT	−	NT
Penechrome	0.1X	−	NT	−	NT	−	NT	−	NT	−	NT	−	NT	−	NT	+	NT	−	NT
17	0.01X	−	NT	−	NT	−	NT	−	NT	−	NT	−	NT	−	NT	−	NT	−	NT
	X	−	NT	−	NT	−	NT	−	NT	−	NT	−	NT	−	NT	−	NT	−	NT
TDL	0.1X	−	NT	−	NT	−	NT	−	NT	−	NT	−	NT	−	NT	−	NT	−	NT
	0.01X	−	NT	−	NT	−	NT	−	NT	−	NT	−	NT	−	NT	−	NT	−	NT
	X	+	C	+	C	+	C	+	C	+	C	+	C	+	C	+	C	+	C
TML	0.1X	+	C	+	C	+	C	+	C	+	C	+	C	+	C	+	C	+	C
	0.01X	+	NT	−	NT	−	NT	−	NT	+	C	+/−	C	−	NT	+	S	−	NT
	X	+	C	+	C	+	C	+	C	+	C	+	C	+	C	+	C	+	C
TQC	0.1X	+	C	+	C	+	C	+	C	+	C	+	C	+	C	+	C	+	C
	0.01X	+	C	+	S	+	C	+	C	+	C	+	C	+	NT	+	C	+	S

K = Knoxville, T = Togus, B = Bloomington, LA = Los Angeles
E = Effectiveness; +, Effective; −, Noneffective; +/−, Slightly Inhibitory
S = Static; C = Biocidal, NT = Not Tested, C- = Contaminated
X,0.1X, and 0.0 1X are concentrations relative to manufacturers' recommended pulse dose.

Work reported by Braun[8] indicated that the use of a biocide containing dithiocarbamates and dithiocarbonate was not effective in controlling the population densities of *Legionella* during the 18 month study period. Further study on the use of tributyltin, a quaternary ammonium biocide, in four evaporative heat exchangers established that these biocides at concentrations recommended by the manufacturers were not effective in controlling *Legionella* in the real world of cooling towers and evaporative condensers.

There is obviously a discrepancy between the laboratory findings and the field data. Part of the reason will include the fact that it is very difficult to maintain a residual in the cooling tower situation that corresponds to the laboratory testing. Gawel and Huddleston[128] have reported that the initial concentration of quaternary ammonium salts of 29 ppm was less than 1 ppm within 3 hr of treatment. The results from a series of tests by Braum[8] indicated that not only was *Legionella* not removed from the tower by the use of 23.7% *n*-alkyl (C14,C16,C12,C18) dimethylbenzyl ammonium chlorides and 2.5% bis (tributyltin) oxide, but another opportunistic pathogen, *Pseudomonas aeruginosa* was also not inhibited by the biocidal treatment. Thus it appears that at the present time the described organic biocidal treatment is not effective in removing *Legionella* from the studied cooling systems.

It is of interest that although *Legionella* is associated with virtually every cooling tower in the country[7,34,92-94] there may well be a set of circumstances that promote the presence and increased density of the organisms. One of these factors may well be the selection of the source water for the various cooling systems. Densities of *Legionella* have been measured in two cooling towers that are identical in construction, biocidal treatment, mode of operation, age, and are within close proximity to one another so that differences between the two towers can be assessed. One tower receives its makeup water from Par Pond, which we have previously shown to contain various serogroups of *L. pneumophila*. The second cooling tower receives makeup water from underground wells which draw from the Tuscaloosa Aquafier at a depth of around 300 feet. We have measured the densities of *Legionella* in each of the towers on a monthly basis since 1978. Although the densities of *Legionella* varied in each tower, the levels of *Legionella* in the tower receiving makeup water from the deep subterranean wells were never as high as the tower which received its makeup water from the surface waters of Par Pond. The data for the various sampling times are shown in Figures 11 and 12, respectively. The densities of *Legionella* in the tower receiving Par Pond water were on occasion higher than the arbitrary levels we have established for treatment.

The only extensive field testing which has demonstrated effectiveness in reducing *Legionella* densities is through the use of chlorine.[92] The technique has been used to reduce the number of *Legionella* in cooling towers as well as in air wash humidifying systems. Our study indicated that free residual levels of chlorine at 1.5 ppm for 72 hr were effective in removing *Legionella* to concentrations that were equivalent to or below source water concentrations. This procedure has instituted chlorination whenever the tower has not received the maintenance dose of chlorine (0.7 to 0.8 ppm for 1 hr each day, free residual) for whatever reason for a period of three consecutive days. This technique has remained effective in our towers. Similar procedures have been instituted at the Medical School of Indiana[136] with similar results. The cooling tower at Burlington, Vermont, was treated with chlorine to a higher residual level with what appears to be similar results but the data were not taken in an ecological fashion.[64]

Simultaneously, data have been collected on the viability of *Legionella* in these systems using the INT methodology during chlorination. The data indicate that *Legionella* is affected by the 1.5 ppm free residual chlorine so that its electron transport system is impaired. The use of the above schedule of chlorination actually lysis the *Legionella* so that the cellular counts of *Legionella* decrease very rapidly with time as does the electron transport system activity. Other biocides do not appear to have this effect. As there may continue to be a

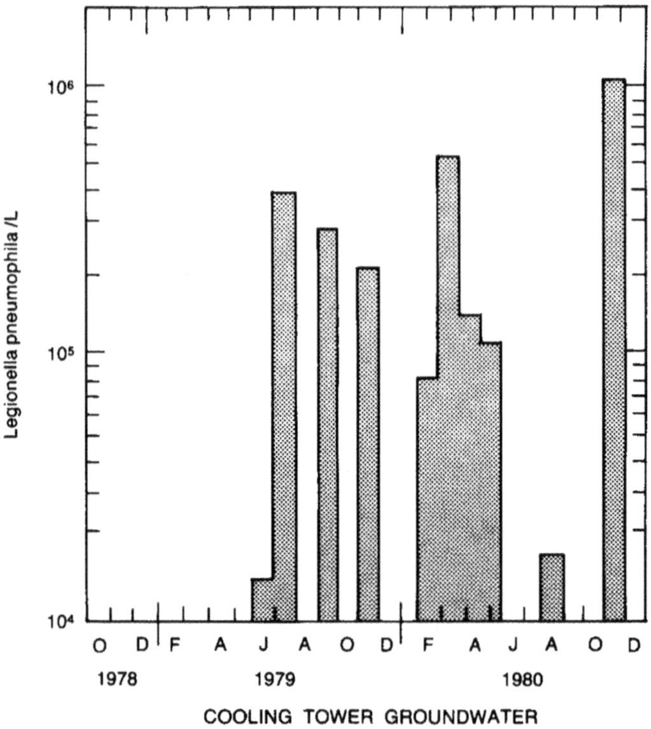

FIGURE 11. Seasonal densities of *L. pneumophila* serogroups 1 to 4 in the basin water of a cooling tower which receives make-up water from underground wells.

question as to whether the DFA is measuring *Legionella* or a similar bacterium that is cross reacting, the chlorination procedure solves this objection very nicely. There is no evidence that indicates that the DFA gives a false negative. That is, whenever *Legionella* is present, it will be stained by the DFA technique. Thus the use of the chlorination procedure removes all the DFA staining organisms and thus assuredly removes the *Legionella*.

The potential drawback to the chlorination procedure is the fact that high levels of free residual chlorine can and will cause corrosion of the cooling tower under nonregulated procedures. The Cooling Tower Institute recommends for redwood towers that the level of continuous chlorination not exceed 1.0 ppm.[129] This is on a continuous free residual basis. The levels recommended here for a short period of time do not appear to cause the problems for corrosion and thus are not harmful to the towers we have treated.

In line with this area of investigation we have monitored a cooling tower that is being treated with the biocide, 1-Bromo-3 chloro-5, 5-dimethylhydantoin, Bromicide™. We have previously shown that the tower contains *Legionella* as initially screened by DFA and subsequently through guinea pig isolations and biochemical techniques. The tower was maintained at recommended Bromicide™ concentrations and the levels of *Legionella* and INT were measured over time as the changes in biocidal concentration was changed. The overall thrust of the study was to determine whether the Bromicide™ treatment would be as effective as the chlorination procedure. The initial thinking is that they would be similar in their effectiveness.

The results of the study are reported in Table 16 . The data show that *Legionella* is relatively unaffected by the concentrations of Bromicide™ up to 2.0 ppm free residual as measured by the free residual halogen methodology described in *Standard Methods 1980*. The data further indicate that not only are the *Legionella* not broken up by the Bromicide™

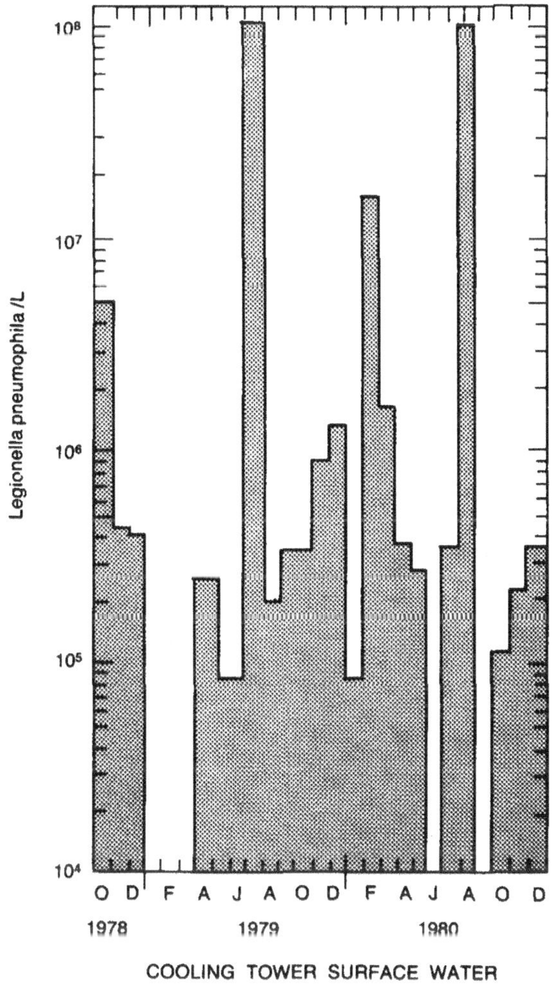

FIGURE 12. Seasonal densities of *L. pneumophila* serogroups 1 to 4 in the basin water of a cooling tower which receives make-up water from Par Pond.

treatment but that the electron transport system is not at all affected by the presence of the Bromicide™. During the study period the levels of *Legionella* never reached concentration where it was necessary to treat the tower by chlorination. It is thus unclear whether the Bromicide™ was effective in keeping the *Legionella* in check through a biostatic process or whether the *Legionella* would not have risen to higher densities in this tower regardless of the biocidal treatment. It is clear that under the situations described *Legionella* is more readily removed from the amplifiers through the use of chlorine than any other of the biocidal treatments either organic or inorganic that have been field tested.

Presently there is no better treatment than the use of chlorine. Obviously from the above discussion and the presented papers any level of chlorination or any regime of chlorination is not appropriate. As we have shown, towers receiving chlorination will have *Legionella* present in the towers. The chlorination procedure will have to be established for each group of towers depending on the source water, the levels of nutrients in the source water, the type of tower as well as the physical-chemical properties of the source water and the cycling times of the water in the tower. All these will play a role in the effectiveness of the biocidal treatment.

Table 16
RESPONSE OF *L. PNEUMOPHILA* IN A COOLING TOWER TREATED WITH THE BIOCIDE, BROMOCIDE™ FROM 4/82 TO 8/82

				L. pneumophila/ℓ			
Sample number	**Temp. (°C)**	**Free halogen**	**% INT positive**	**Knoxville**	**Togus**	**Bloomington**	**Los Angeles**
46-1	21.3	0.1 ppm	48	8.46×10^5	2.07×10^6	2.46×10^5	3.82×10^5
46-2	24.1	0.2 ppm	NT	1.64×10^5	3.74×10^6	2.40×10^5	3.82×10^5
46-3	22.6	0.5 ppm	53	2.24×10^5	2.24×10^6	2.10×10^5	1.64×10^5
46-4	23.1	0.6 ppm	NT	2.73×10^4	2.62×10^4	8.19×10^4	3.82×10^5
46-5	24.4	0.6 ppm	57	5.46×10^4	2.62×10^6	8.19×10^4	2.18×10^5
49-6	24.1	0.9 ppm	NT	5.46×10^4	2.95×10^6	5.46×10^4	1.64×10^5
46-7	24.3	1.4 ppm	63	3.47×10^4	3.47×10^6	5.46×10^4	4.37×10^5
46-8	24.7	1.3 ppm	51	3.47×10^4	3.52×10^6	5.46×10^4	4.37×10^5
47-1	18.7	0.8 ppm	52	2.46×10^5	2.51×10^6	8.21×10^4	2.46×10^5
47-2	18.9	0.8 ppm	48	4.69×10^5	1.09×10^6	6.09×10^5	3.00×10^5
47-3	15.8	0.9 ppm	52	5.42×10^4	1.12×10^6	5.46×10^4	1.64×10^5
[a]47-4	20.6	0.0 ppm	43	1.91×10^5	1.91×10^5	9.10×10^3	BD
52-1	19.1	1.9 ppm	49	1.30×10^5	1.20×10^6	BD	3.55×10^5
52-2	21.9	1.5 ppm	44	2.18×10^5	1.10×10^6	BD	1.09×10^5
52-3	19-7	0.4 ppm	NT	8.19×10^5	5.19×10^5	BD	5.46×10^4
53-1	20.4	2.1 ppm	49	1.09×10^5	9.01×10^5	BD	1.64×10^5
53-2	19.3	0.7 ppm	63	8.74×10^5	2.07×10^6	BD	1.09×10^5
54-1	19.8	0.8 ppm	54	4.91×10^5	2.10×10^6	BD	1.09×10^5
55-1	23.4	0.4 ppm	48	4.10×10^5	3.99×10^6	BD	1.45×10^6
58-7	25.5	0.4 ppm	51	1.91×10^5	5.81×10^6	BD	1.45×10^6
61-3	25-9	0.3 ppm	50	1.85×10^5	6.83×10^6	BD	1.37×10^5
64-16	26-3	0.9 ppm	83	5.19×10^5	1.37×10^6	1.09×10^5	3.22×10^6

[a] Make-up water; NT = Not Tested; BD = Below Detection.

It must be remembered that *Legionella* is an aquatic bacterium so that any system using water is likely to be a recipient of *Legionella* and the treatment of a system for the removal of *Legionella* will not be a one time event. Cooling systems should be monitored for *Legionella* on a regular basis in the summer time particularly mid to late summer and early fall. This will depend upon the juxtaposition of the amplifier to the population at large. The corresponding tower should be monitored again in the spring of the year when the tower is just being brought into operation. The frequency of the monitoring will depend a large part of the effectiveness of the biocidal treatment and the source of the make-up water.

At the present time no surrogate exists for measuring *Legionella*. Clean looking towers or those without slime are not an indication that they contain low levels of *Legionella*. On the other hand, towers that are extremely fouled with slime and high densities of microbial growth may have densities of *Legionella* that are very low. The converse for each of these systems is equally true. There is simply no way at present to tell whether or not a particular system is relatively free or highly contaminated with *Legionella* without doing the proper tests which have been described.

VII. RESEARCH PERSPECTIVES

I believe that any review of this sort would be incomplete without a research perspective. As we all know, the more questions one answers the more knowledge is gained and the more questions are presented that need answered. *Legionella* ecology is no exception.

Certainly a great deal needs to be learned about the virulence changes of *Legionella* in the natural habitat and what turns the bacterium on and off. We have seen a cycle of virulence in the natural habitat and that cycle has been repeatable for the infection of guinea pigs.

There have been a number of studies on the changes of virulence in the laboratory setting, but such information is of limited value for the environmental setting where *Legionella* finds its home.

The whole question of iron and its role in the presence of infection with regard to *Legionella* needs to be investigated. Such studies may provide a surrogate for *Legionella* in the natural environment through the screening of iron compounds or their various states of activity.

A very important area that should receive top priority in the ecological research of *Legionella* is the area of antibiotic resistance and plasmids. It has been shown that some of the *Legionella* have plasmids within their genetic make-up.[130,131] Generally the antibiotic resistance is carried on the plasmid material. Orrison et al.[9] have demonstrated that various species of *Legionella* differ in their ability to respond in vivo to various antibiotics. That is, *L. pneumophila* is quite susceptible to erythromycin, rifampin, and cephalothin while *L. bozemanii* and *L. micdadei* are more resistent to erythromycin, the drug of choice for *Legionella* infection. The habitats that foster and amplify *L. pneumophila* appear to be the same habitats that the other species of *Legionella* reside in. Thus, large population densities may provide the necessary means for the exchange of genetic material between the various species of *Legionella* and the potential production of a *Legionella* resistant to a wide range of antibiotics.

The question of interactions in the environment with other components of the environment has not been fully explored. Though we have seen that *Legionella* is indeed an aquatic bacterium, it is not clear whether an association with any terrestrial habitats exists. Such habitats, if present, could provide inoculum to the aquatic environment.

The question of aerosolization points out the fact that many of the nation's power generating facilities are not once through systems so that they are indeed dependent on some form of cooling lake or cooling towers. Because of the large amounts of cooling water used in these systems the cost of biocidal treatment is almost out of the question, yet the cost needs to be balanced against the impact to the worker and to the surrounding community. The dissemination of *Legionella* away from these various point sources needs to be determined and fully evaluated in terms of modeling and verifications of those models. To the credit of the Electric Power Research Institute they are aware, responsive, and funding some of the above questions.[134]

The establishing of guidelines for the operation of cooling systems and the proper disinfection of water for amplifiers needs to be addressed. Such guidelines can only be established with the proper research placed in the hands for legislative processes.

Knowledge demands responsibility and the scientist is much to blame in the ineffectual use of that knowledge. Knowledge must be communicated in truth and in the perspective that man is indeed a special being who takes his highest place in creation.

ACKNOWLEDGMENTS

I thank W. B. Cherry and R. L. Tyndall for their continued encouragement and insight into the ecological fascinations of *Legionella*. My research has been supported by National Institutes of Health grant 1-RO1-AI17673-01 and by the contract no. DE-AC09-76SR00001 with the Department of Energy.

REFERENCES

1. **Blackmon, J. A., Chandler, F. W., Cherry, W. B., England, A. C., Feeley, J. C., Hicklin, M. D., McKinney , R. M., and Wilkinson, H. Q.,** Legionellosis, *Am. J. Pathol.*, 103, 429, 1981.

2. **Cordes, L. G. and Fraser, D. W.,** Legionellosis. Legionnaires' disease; Pontiac fever, *Med. Clin. N. Am.*, 64, 395, 1980.
3. **Band, J. D. and Fraser, D. W.,** Legionellosis (Legionnaires' Disease and Pontiac Fever), in *Medical Microbiology and Infectious Diseases*, Braude, A. I., Ed., W. B. Saunders, Philadelphia, 1981.
4. **Fliermans, C. B., Cherry, W. B., Tison, D. L., Smith, R. B., and Pope, D. H.,** Distribution of *Legionella pneumophilia, Appl. Environ. Microbiol.*, 41, 9, 1981.
5. **Fliermans, C. B., Cherry, W. B., Orrison, L. H., and Thacker, L.,** Isolation of *Legionella pneumophila* from nonepidemic-related aquatic habitats, *Appl. Environ. Microbiol.*, 37, 1239, 1979.
6. **Christensen, S. W., Tyndall, R. L., Soloman, J. A., Fliermans, C. B., and Gough, S. B., Eds.,** Legionnaires' Disease Bacterium in Power Plant Cooling Systems, Electric Power Research Institute, Palo Alto, Calif., 1982, 97.
7. **Orrison, L. H., Cherry, W. B., and Milan, D.,** Isolation of *Legionella pneumophila* from cooling tower by filtration, *Appl. Environ. Microbiol.*, 41, 1202, 1981.
8. **Braun, E. B.,** Factors Affecting the Distribution, Seasonal Response and Control of *Legionella pneumophila* in Aquatic Enviroments, Ph.D. thesis, Rensselaer Polytechnic Institute, Troy, N.Y., 1982.
9. **Orrison, L. H., Cherry, W. B., Fliermans, C. B., Dees, S. B., McDougal, L. K., and Dodd, D. J.,** Characteristics of environmental isolates of *Legionella pneumophila. Appl. Environ. Microbiol.*, 42, 109, 1981.
10. **Fliermans, C. B., Gorden, R. W., Hazen, T. C., and Esch, G. W.,** *Aeromonas* distribution and survival in a thermally altered lake, *Appl. Environ. Microbiol.*, 33, 114, 1977.
11. **Tansey, M. R. and Fliermans, C. B.,** Pathogenic Species of Thermophilic and Thermotolerant Fungi in Reactor Effluents of the Savannah River Plant, Rep. CONF-77114, U. S. Department of Energy, Washington, D. C., 1978
12. **Hazen, T. C.,** Ecology of *Aeromonas hydrophila* in a South Carolina cooling reservoir, *Microb. Ecol.* 5, 179, 1979.
13. **Bopp, C. A., Sumner, J. W., Morris, G. K., and Wells, J. G.,** Isolation of *Legionella* spp. from environmental water samples by low-pH treatment and use of a selective medium, *J. Clin. Microbiol.*, 13, 714, 1981.
14. **Cordes, L. G., Fraser, D. W., Skaliy, P., Perlino, C. A., Elsea, W. R., Mallison, G. F., and Hayes, P. S.,** Legionnaires' disease outbreak at an Atlanta, Georgia, country club: evidence for spread from an evaporative condenser, *Am. J. Epidemiol.*, 111, 425, 1980.
15. **Edelstein, P. H., Snitzer, J. B., and Finegold, S. M.,** Isolation of *Legionella pneumophila* from hospital potable water specimens: comparison of direct plating with guinea pig inoculations, *J. Clin. Microbiol.*, 15, 1092, 1982.
16. **Wadowsky, R. M., Yee, R. B., Mezmar, L., Wing, E. J., and Dowling, J. N.,** Hot water systems as sources of *Legionella pneumophila* in hospital and nonhospital plumbing fixtures, *Appl. Environ. Microbiol.*, 43, 1104, 1982.
17. **Wadowsky, R. M. and Yee, R. B.,** A glycine-containing selective medium for isolation of *Legionellacae* from environmental specimens, *Appl. Environ. Microbiol.*, 42, 768, 1981.
18. **Feeley, J. C., Gibson, R. J., Gorman, G. W., Langford, N. C., Rasheed, J. K., Mackel, D. C., and Baine, W. B.,** Charcoal-yeast extract agar: primary isolation medium for *Legionella pneumophila, J. Clin. Microbiol.*, 10, 437, 1979.
19. **Thorpe, T. C. and Miller, R. D.,** Negative enrichment for the isolation of *Legionella pneumophila* from seeded cooling tower water, *Appl. Environ. Microbiol.*, 40, 849, 1980.
20. **Pasculle, A. W., Feeley, J. C., Gibson, R. J., Cordes, L. G., Meyerowitz, R. L., Patton, C. M., Gorman, G. W., Carmack, C. L., Ezzell, J. W., and Dowling, J. N.,** Pittsburgh pneumonia agent: direct isolation from human lung tissue, *J. Infect. Dis.*, 141, 727, 1980.
21. **Vickers, R. M., Brown, A., and Garrity, G. M.,** Dye-containing buffered charcoal-yeast extract medium for differentiation of members of the family *Legionellaceae, J. Clin. Microbiol.*, 13, 380, 1981.
22. **Hendry, G. R.,** Aquatic Task Force on Environmental Assessment of the Atikokan Power Plant: Effects on Aquatic Organisms, Rep. BNL 50932, U. S. Department of Energy, Washington, D. C., 1978.
23. **Satake, K. and Saijo, Y.,** Carbon dioxide content and metabolic activity of microorganisms in some acid lakes in Japan, *Limnol. Oceanogr.*, 19, 331, 1974.
24. **Dugan, P. R., MacMillan, C. B., and Pfister, R. M.,** Aerobic heterotrophic bacteria indigenous to pH 2.8 acid mine water: predominant slime-producing bacteria in acid streamers, *J. Bacteriol.*, 101, 982, 1970.
25. **Brock, T. D.,** *Thermophilic Microorganisms and Life at High Temperatures*, Springer-Verlag, New York, 1978, 465.
26. **Tison, D. L., Pope, D. H., Cherry, W. B., and Fliermans, C. B.,** Growth of *Legionella pneumophila* in association with blue-green algae (Cyanobacteria), *Appl. Environ. Microbiol.*, 39, 456, 1980.
27. **Fraser, D. W. and McDade, J. E.,** Legionellosis, *Sci. A.*, 241, 82, 1979.
28. **McDade, J. E. and Shepard, C. C.,** Virulent to avirulent conversions of Legionnaires' Disease bacterium *(Legionella pneumophila* — its effect on isolation techniques, *J. Infect. Dis.*, 139, 707, 1979.

29. **Elliott, J. A. and Johnson, W.,** Virulence conversion of *Legionella pneumophilia* serogroup 1 by passage in guinea pigs and embryonated eggs, *Infect. Immun.*, 35, 943 , 1982.
30. **Fliermans, C. B., Soracco, R. J., and Pope, D. H.,** Measure of *Legionella pneumophila* activity *in situ*, *Curr. Microbiol.*, 6, 89, 1981.
31. **Hazen, T. C. and Fliermans, C. B.,** The distribution of *Aeromonas hydrophila* in natural and man-made habitats, *Appl. Environ. Microbiol.*, 38, 166, 1979.
32. **Tison, D. W., Wilde, E. W., Pope, D. H., and Fliermans, C. B.,** Productivity and species composition of algal mat communities exposed to a fluctuating thermal regime, *Microb. Ecol.*, 7, 151, 1981.
33. **Stout, J., Yu, V. L., Vickers, R. M., Zuravleff, M. D., Brown, A., Yee, R. B., and Wadowsky, R.,** Ubiquitousness of *Legionella pneumophila* in the water supply of a hospital with endemic Legionnaires' Disease, *N. Engl. J. Med.* 303, 466, 1982.
34. **Tyndall, R. J.,** Concentrations, serotypic profiles and infectivity of Legionnaires' Disease bacteria populations in cooling tower, *Cooling Tower Inst.*, 3, 25, 1982.
35. **Tyndall, R. L., Fliermans, C. B., Gough, S. B., Domingue, E., and Duncan, C.,** Isolation of a new *Legionella* species from thermally altered waters, *Curr. Microbiol.*, 9, 77, 1983.
36. **Cherry, W. B., Pittman, B., Harris, P. P. , Herbert, G. A., Thomason, B. M., Thacker, L., and Weaver, R. E .,** Detection of Legionnaires' Disease bacterium by direct immunofluorescent staining, *J. Clin. Microbiol.*, 8, 329, 1978.
37. **Orrison, L. H., Bibb, W. F., and Cherry, W. B.,** Antigenic relationship among Legionellae and non-Legionellae by direct fluorescent antibody and immunodiffusion studies, *Abstr. Annu. Meet. Soc. Microbiol.*, 82, 278, 1982.
38. **Moss, C. W., Weaver, R. E., Dees, S. B., and Cherry, W. G.,** Cellular fatty acids composition of isolates from Legionnaires' Disease, *J. Clin. Microbiol.*, 6, 149, 1977.
39. **Brenner, D. J., Steigerwalt, G. A., Weaver, R. E., McDade, J. E., Feeley, J. C., and Mandell, M.,** Classification of the Legionnaires' Disease bacterium: an interim report, *Curr. Microbiol.*, 1, 71, 1979.
40. **Coons, A. H., Chreech, J. F., Jones, R. N., and Berliner, E.,** Demonstration of pneumococcal antigen in tissues by the use of fluorescent antibody, *J. Immunol.*, 45, 159, 1942.
41. **Schmidt, E. L.** Fluorescent antibody techniques for the study of microbial ecology, *Bull. Ecol. Res. Commun.*, 17, 67, 1973.
42. **Schmidt, E. L. and Bankole, R. O.,** Specificity of immunofluorescent staining for study of *Aspergillus flavus* in soil, *Appl. Microbiol.*, 13, 673, 1965.
43. **Fliermans, C. B. and Schmidt, E. L.,** Fluorescence microscopy: direct detection, enumeration, and spatial distribution of bacteria in aquatic systems, *Arch. Hydrobiol.*, 76, 33, 1975.
44. **Henry, R. A., Johnson, R. C., Bohlool, B. B., and Schmidt, E. L.,** Detection of *Leptospira* in soil and water by immunofluorescent staining, *Appl. Microbiol.*, 21, 953, 1971.
45. **Fliermans, C. B. and Schmidt, E. L.,** Immunofluorescence for autecological study of a unicellular blue-green alga, *J. Phycol.*, 13, 364, 1977.
46. **Gordon, R. W., Hazen, T. C., and Fliermans, C. B.,** The presence of *Aeromonas hydrophila* associated with *Alligator mississippiensis, Wildl. Dis.*, 15, 239, 1979.
47. **Hazen, T. C., Fliermans, C. B., Hirsh, R. P., Esch, G. W.,** Prevalence and distribution of *Aeromonas hydrophila* in the United States, *Appl. Environ. Microbiol.*, 34, 731, 1978.
48. **Fliermans, C. B., Tyndall, R. L., Domingue, E. L., and Willaert, E. J. P.,** Isolation of *Naegleria fowleri* from artificially heated water, *J. Thermal Biol.*, 4, 303, 1979.
49. **Wilde, E. W. and Fliermans, C. B.,** Fluorescence microscopy for algal studies, *Trans. Am. Microsc. Soc.*, 98, 96, 1979.
50. **Cherry, W. B., Pittman, B., Harris, P. P., Herbert, G. A., Thomason, B. M., and McKinney, R. M.,** Legionnaires' Disease—the disease and the bacterium with special reference to the laboratory diagnosis, in *Immunofluorescence and Related Staining Techniques*, Knapp, W., Holubar, K., and Wick, G., Eds., Elsevier/North Holland Biomedical Press, New York, 1978, 163.
51. **Brock, T. D.,** Use of fluorescence microscopy for quantifying phytoplankton, especially filamentous blue-green algae, *Limnol. Oceanogr.*, 23, 158, 1978.
52. **Cherry, W. B. and McKinney, R. M.,** *Legionella*, mycobacteria, and Freunds' adjuvant, *N. Engl. J. Med.* , 8, 1242, 1979.
53. **Katz, S. M. and Hammel, J.,** Confusion of *Legionella pneumophila* and *Mycobacterium tuberculosis, N. Engl. J. Med.*, 300, 1490, 1979.
54. **Tison, D. L . and Wilde, E. W.,** Primary production and biovolume of various phototrophic plankton size fractions in three southeastern United States reservoirs, *Appl. Environ. Microbiol.*, 41, 1055, 1981.
55. **Fliermans, C. B., Schenider-Cain, P., and Schmidt, E. L.,** Direct measurement of bacterial stratification in Minnesota lakes, *Arch. Hydrobiol.*, 76, 248, 1975.
56. **Packard, T. T., Healy, M. L., and Richards, F. A.,** Vertical distribution of the activity of the respiratory electron transport system in marine plankton, *Limnol. Oceanogr.*, 16, 60, 1971.

57. **Zimmerman, R., Iturriaga, R., and Becker-Brick, J.** , Simultaneous determination of the total number of aquatic bacteria and the number thereof involved in respiration, *Appl. Environ. Microbiol.*, 36, 926, 1978.
58. **Packard, T. T.,** The measurement of respiratory electrontransport activity in marine phytoplankton, *J. Marine Res.*, 29, 235, 1971.
59. **Winn, W. C., Jr. and Chandler, F. W.,** Role of virulence factors in *Legionella* infections, *Arch. Pathol. Lab. Med.*, 106, 105, 1982.
60. **Marshall, K. D.,** *Interfaces in Microbial Ecology*, Harvard University Press, Cambridge, Mass., 1976, 156.
61. **Hazen, T. C.,** The Ecology of *Aeromonas hydrophila* in a South Carolina Cooling Reservoir, Ph.D. thesis, Wake Forest University, Winston Salem, N. C., 1978.
62. **Tobin, J. O., Beare, J., Dunhill, M. S., Fisher-Hock, S., French, M., Mitchell, R. G. , Morris, P. J., and Muers, M. F.,** Legionnaires' disease in a transplant unit: isolation of the causative agent from shower baths, *Lancet*, 2, 118, 1980.
63. **Dondero, T. J., Jr., Rendtorff, R. C., Mallison, G. F., Weeks, R. M., Levy, J. S., Wong, E. W., and Schaffner, W.,** An outbreak of Legionnaires' Disease associated with a contaminated air conditioning cooling tower, *N. Engl. J. Med.*, 302, 365, 1980.
64. **Marshal, E.,** Hospitals harbor a built-in disease source, *Science*, 210, 145, 1980.
65. **Stout, J., Vickers, R. M., Yu, V. L., and Shonnard, J.,** Potable water supply as the hospital reservoir for Pittsburgh pneumonia agent, *Lancet*, 1, 471, 1982.
66. **Cordes, L. G., Wiesenthal, A. M., Gorman, G. W., Phair, J. P., Sommers, H. M., Brown, A., Yu, V. L., Magnussen, M. H., Mayer, R. D., Wolf, J. S., Shands, K. N., and Fraser, D. W.,** Isolation of *Legionella pneumophila* from hospital shower heads, *Ann. Intern. Med.*, 94, 195, 1980.
67. **Miller, R. P.,** Cooling towers and evaporative condensers, *Ann. Intern. Med.* 90, 667, 1979.
68. **Brown, A., Yu, V. L., Magnussen, M. H., Vickers, R. M., Garrity, G. M., and Elder, E. M.,** Isolation of Pittsburgh agent from a hospital shower, *Appl. Environ. Microbiol.*, 43, 725, 1982.
69. **Edelstein, P. H., Snitzer, J. B., Finegold, S. M.,** Isolation of *Legionella pneumophila* from hospital water specimens: comparison of direct plating with guinea pig inoculations, *J. of Clin. Microbiol.*, 15, 1092, 1982.
70. **Tobin, J. O. H., Bartlett, C. L. R., Watkins, S. A. , Barrow, G. L., Macrae, A. D., Taylor, A. G., Fallon, R. J., and Lynch, F. R. N.,** Legionnaires' Disease: further evidence to implicate water storage and distribution systems as sources, *Br. Med. J.*, 282, 573, 1981.
71. **Tobin, J. O. H., Swann, R. A., and Bartlett, C. L. R.,** Isolation of *Legionella pneumophila* from water systems: methods and preliminary results, *Br. Med. J.*, 282, 515, 1981.
72. **Dournon, E., Bure, A., Desplaces, N., Carette, M. F., and Mayaud, Ch.,** Legionnaires' Disease related to gastric lavage with tap water, *Lancet*, 1, 797, 1982.
73. **Cherry, W. B., Gorman, G. W., Orrison, L. H., Moss, W. C., Steigerwalt, A. G., Wilkinson, H. W., Johnson, S. C., McKinney, R. M., and Brenner, D. J.,** *Legionella jordanis*: a new species of *Legionella* isolated from water and sewage, *J. Clin. Microbiol.*, 15, 295, 1982.
74. **Yee, R. B. and Wadowsky, R. M.,** Multiplication of *Legionella pneumophila* in unsterilized tap water, *Appl. Environ. Microbiol.*, 43, 1330, 1982.
75. **Lewis, W. M.,** Evaluation of heat distribution in a South Carolina reservoir receiving heated water, in *Thermal Ecology*, Gibbons, J. W. and Sharitz, R. R., Eds., Atomic Energy Commission Series, CONF-730505, Washington, D. C., 1974.
76. **Gibbons, J. W. and Sharitz, R. R.,** Thermal alteration of aquatic ecosystems, *Am. Sci.*, 62, 660, 1974.
77. **Tilly, L. J.,** Changes in water chemistry and primary productivity of a reactor cooling reservoir (Par Pond), in Mineral Cycling in the Southeastern Ecosystems, Howell, F. G., Gentry, J. B., and Smith, M. M., Eds., Energy Research and Development Administration Symposium Series, CONF-740513, Washington, D. C., 1975.
78. **Neill, J. S. and Babcock, D. F.,** The Dissipation of Reactor Heat at the Savannah River Plant, United States Department of Energy, DP-1247, E. I.Du Pont de Nemours and Co., Wilmington, Del., 1971.
79. **Heinen, W.,** Extreme thermophilic bacteria: fatty acids and pigments, *Antonie van Leeuwenhoek; J. Microbiol. Serol.*, 36, 582, 1970.
80. **Shen, P. Y., Coles, E., Foote, J. L., and Stenesh, J.,** Fatty acid distribution in the mesophilic and the thermophilic strains of the genus *Bacillus*, *J. Bacteriol.*, 103, 479, 1971.
81. **Weerkamp, A. and Heinen, W.,** Effects of temperature on the fatty acid composition of the extreme thermophiles *Bacillus caldolyticus* and *Bacillus caldotenax*, *J. Bacteriol.*, 109, 443, 1972.
82. **Gorden, R. W. and Fliermans, C. B.,** Survival and viability of *Escherichia coli* in a thermally altered reservoir, *Water Res.*, 12, 343, 1978.
83. **Kasweck, K. L. and Fliermans, C. B.,** Lactose variability of *Escherichia coli* in thermally stressed reactor effluent waters, *Appl. Environ. Microbiol.*, 36, 739, 1978.

84. **Ortiz-Roque, C. and Hazen, T. C.,** Isolation of *Legionella pneumophila* from tropical waters, *Abstr. Annu. Meet. Am. Soc. Microbiol.*, 82, 180, 1982.
85. **Tison, D. W., Baross, J. A., and Seidler, R. J.,** *Legionella* in aquatic habitats in the Mt. St. Helens blast zone, *Abstr. Annu. Meet. Am. Soc. Microbiol.*, 82, 212, 1982.
86. **Thacker, S. B., Bennett, J. V., Tsai, T. F., Fraser, D. W., McDade, J. E., Shepard, C. C., Williams, K. H., Fr., Stuart, W. H., Dull, H. B., and Eickoff, T. C.,** An outbreak in 1965 of severe respiratory illness caused by the Legionnaires' Disease bacterium, *J. Infect. Dis.*, 138, 512, 1978.
87. **Cordes, L. G. and Fraser, D. W.,** Legionellosis: Legionnaires' Disease; Pontiac fever, *Med. Clin. N. Am.*, 64, 395, 1980.
88. **Imperato, P. J.,** Legionellosis and the indoor environment, *Bull. N. Y. Acad. Med.*, 57, 922, 1981.
89. **Brock, T. D. and Yoder, I.,** Thermal pollution of a small river by a large university: bacteriological studies. *Proc. Indiana Acad. Sci. 1970*, 80, 183, 1971.
90. American Society of Heating, Refrigeration and Air Conditioning Engineers, Position Paper on Legionellosis, ASHRAE, New York, 1981.
91. **Adams, A. P. and Lewis, B. G.,** Microbial Aerosols from Cooling Towers and Cooling Sprays: A Pilot Study, NRC FIN No . A2098, United States Nuclear Regulatory Commission, Washington, D. C., 1979.
92. **Fliermans, C. B., Bettinger, G. E., and Fynsk, A. W.,** Treatment of cooling systems containing high levels of *Legionella pneumophila, Water Res.*, 16, 903, 1982.
93. **Bettinger, G. E., Fynsk, A. E., and Fliermans, C. B.,** Detection of *Legionella pneumophila* in cooling towers not associated with disease outbreak, *Abstr. Annu. Meet. Am. Soc. Microbiol.*, 80, 288, 1980.
94. **England, A. C., Fraser, D. W., Mallison, G. F., Mackel, D. C., Skaliy, P., and Gorman, G. W.,** Failure of *Legionella pneumophila* sensitivitics to predict culture results from disinfectant-treated air-conditioning cooling towers, *Appl. Environ. Microbiol.*, 43, 240, 1982.
95. **Berendt, R. F., Young, H. W., Allen, R. G., and Knutsen, G. L.,** Dose-response of guinea pigs experimentally infected with aerosols of *Legionella pneumophila, J. Infect. Dis.*, 141, 186, 1980.
96. **Cherry, W. B. and McKinney, R. M.,** Detection in clinical specimens by direct immunofluorescence, in *Legionnaires- The Disease, the Bacterium and Methodology*, Jones, G. L. and Hebert, G. A., Eds., Center for Disease Control, Atlanta, Ga.,1978.
97. **Girod, J. C., Reichman, R. C., Winn, W. C., Jr., Klaucke, D. N., Vogt, R. L. , and Dolin, R.,** Pneumonic and non-pneumonic forms of Legionellosis, *Arch. Intern. Med.*, 142, 545, 1982.
98. **Bakerville, A., Fitzgeorge, R. B., Broster, M., Hambleton, P., and Dennis, P. J.,** Experimental transmission of Legionnaires' Disease by exposure to aerosols of *Legionella pneumophila, Lancet*, 2, 1389, 1981.
99. Center for Disease Control, Legionnaires' Disease—United States, Morbidity and Mortality Weekly Report, 27, 439, 1978.
100. **Tison, D. W.,** Primary Productivity, Heterotrophy, Metabolic Indicators of Stress and Interactions in Algal-Bacterial Mat Communities Affected by a Fluctuating Thermal Regime, Ph.D. thesis, Rensselaer Polytechnic Institute, Troy, N.Y., 1980.
101. **Berendt, R. F.,** Influence of blue-green algae (*Cyanobacteria*) on survival of *Legionella pneumophila* in aerosols, *Infect. Immun.*, 32, 690, 1981.
102. **Tison, D. W. and Lingg, A. J.,** Dissolved organic carbon utilization and O_2 uptake in algal-bacterial microcosms, *Can. J. Microbiol.*, 25, 1315, 1979.
103. **Caldwell, D. W. and Caldwell, S. J.,** A *Zoogleа* sp. associated with blooms of *Anabaena flosaquae, Am. J. Microbiol.*, 24, 922, 1978.
104. **Lange, W.,** Effects of carbohydrates in the symbiotic growth of planktonic blue-green algae with bacteria, *Nature (London)*, 215, 1277, 1967.
105. **Paerl, H. W.,** Specific associations of the blue green algae *Anabaena* and *Aphanizomenon* with bacteria in fresh water blooms, *J. Phycol.*, 12, 431, 1976.
106. **Wiebe, W. J. and Smith, D. F.,** Direct measurement of DOC released by phytoplankton and incorporation by microheterotrophs, *Marine Biol.*, 42, 213, 1977.
107. **Vogel, S. L., Fritch, H. L., and Gotham, I. J.,** Qualitative assay of dissolved amino acids and sugars excreted by *Chlamydomonas reinhardtii* (*Chlorophyceae*) and *Euglena gracilis* (*Euglenophyceae*), *J. Phycol.*, 14, 403, 1978.
108. **Fogg, G. E.,** Extracellular products, in *Physiology and Biochemistry of Algae* , Lewin, L. A., Ed., Academic Press, New York, 1962.
109. **Hellebust, J. A.,** Extracellular products, in *Algal Physiology and Biochemistry*, Stewart, W. P. O., Ed., University of California Press, Berkeley, Calif., 1974.
110. **Kuentzel, L. E.,** Bacteria, carbon dioxide and algal blooms *J. Water Pollut. Control Fed.*, 41, 1737, 1969.
111. **Lange, W.,** Enhancement of algal growth in Cyanophytabacterial systems by carbonaceous compounds, *Can. J. Microbiol.*, 17, 303, 1971.

112. **McFeeters, G., Stewart, S. A., and Olson, S. B.,** Growth of heterotrophic bacteria and algal extracellular products in oligotrophic waters, *Appl. Environ. Microbiol.*, 35, 383, 1978.
113. **Reid, G. K.,** *Ecology of Inland Waters and Estuaries*, Reinhold, New York, 1965, 375.
114. **Grace, J. B. and Tilly, L. J.,** Distribution and abundance of submerged macrophytes, including *Myriophyllum spicatum* L. (Angiospermae), in a reactor cooling reservoir, *Arch. Hydrobiol.*, 77, 475, 1976.
115. **Stanley, R. A.,** Studies on Nutrition, Photosynthesis and Respiration in *Myriophyllum spicatum* L., Ph. D. thesis, Duke University, Durham, N. C., 1970.
116. **Tyndall, R. L., Domingue, E. L., Willaert, E., and Stevens, A. R.,** Presence of pathogenic *Naegleria* in power plant cooling lakes, *Abstr. Annu. Meet. Am. Soc. Microbiol.*, 79, 231, 1979.
117. **Fliermans, C. B., Tyndall, R. L., Domingue, E. L., and Willaert, E. J. F.,** Isolation of *Naegleria fowleri* from artificially heated water, *J. Thermal Biol.*, 4, 303, 1979.
118. **Robotham, T. J.,** Pontiac fever explained? *Lancet*, 2, 969, 1980.
119. **Collins, M. T., Cho, S. N., and Reif, J. S.,** Prevalence of antibodies to *Legionella pneumophila* in animal populations, *J. Clin. Microbiol.*, 15, 130, 1982.
120. **McFeters, G. A. and Stuart, D. G.,** Survival of coliform bacteria in natural waters: field and laboratory studies with membrane-filter chambers, *Appl. Microbiol.*, 24, 805, 1972.
121. **Fliermans, C. B. and Gorden, R. W.,** Modification of membrane diffusion chambers for deep water studies. *Appl. Environ. Microbiol.*, 33, 207, 1977.
122. **Fliermans, C. B.,** Growth of *Legionella pneumophila* in diffusion chambers, *Abstr. Annu. Meet. Am. Soc. Microbiol.*, 82, 180, 1982.
123. **Grace, R. D., Dewar, N. E., Barnes, W. G., and Hodges, G. R.,** Susceptibility of *Legionella pneumophila* to three cooling tower microbiocides, *Appl. Environ. Microbiol.*, 41, 233, 1981.
124. **Skaliy, P., Thompson, T. A., Gorman, G. W., Morris, G. K., McEachern, H. V., and Machel, D. C.,** Laboratory studies of disinfectants against *Legionella pneumophila, Appl. Environ. Microbiol.*, 40, 697, 1980.
125. **Soracco, R. J.,** Effectiveness of Cooling Tower Biocides against Algae and *Legionella pneumophila* and Mode of Action of Bis(tributyltin) Oxide, Ph.D. thesis, Rensselaer Polytechnic Institute, Troy, N.Y., 1981.
126. **Wang, W. L., Blaser, M. J., Cravens, J., and Johnson, M. A.,** Growth, survival and resistance of the Legionnaires' Disease bacterium *Ann. Intern. Med.*, 90, 614, 1978.
127. **Zedler, R. J.,** Organotins as industrial biochemicals, *Tin and Its Uses*, 53, 7, 1961.
128. **Gawel, L. J. and Huddleston, R. L.,** The Biodegradability of Low Concentrations of Certain Quarternary Ammonium Antimicrobials by Bacteria, paper presented at the American Oil Chemists National Meeting, Los Angeles, Calif., 1972.
129. **White, G. C.,** *Handbook of Chlorination for Potable Water, Waste Water, Cooling Water, Industrial Processes and Swimming Pools*, Van Nostrand Reinhold, New York, 1972.
130. **Johnson, S. R. and Schalla, W. O.,** Plasmids of serogroup 1 strains of *Legionella pneumophila, Cur. Microbiol.*, 7, 143, 1982.
131. **Aye, T., Wachsmuth, K., Feeley, J. C., Gibson, R. J., and Johnson, S. R.,** Plasmid profiles of *Legionella* species, *Curr. Microbiol.*, 6, 389, 1981.
132. **Fliermans, C. B.,** unpublished results.
133. **Pope, D. H. and Fliermans, C. B.,** unpublished results.
134. **Tyndall, R. L., Fliermans, C. B., Gough, S. B., and Christensen, S. W.,** Legionnaries Disease Bacterium in Power Plant Cooling Systems: Phase I Final Report, Electric Power Research Institute, Palo Alto, Cal., EPRI EA-3153, 1983.
135. **Cherry, W. B., Fliermans, C. B., and Tiedje, J. M.,** unpublished results.
136. **French, Z. Q,** personal communication.
137. **Tyndall, R. L. and Dominque, E. L.,** Cocultivation of *Legionella pneumophilia* and free living amoebae, *Appl. Environ, Microbiol.*, 44, 954, 1983.

Part III: Pathobiology of Legionellosis

Chapter 7

ANIMAL MODELS OF LEGIONELLOSIS

Sheila Moriber Katz and Joseph P. Matus

TABLE OF CONTENTS

I. INTRODUCTION

Animal models are invaluable systems for investigation of infectious agents. They are amenable to manipulations of variables not possible in man because of ethical restraints. The major requirement of any model is that it mimic human form of the disease, enough so that the findings of the model are applicable to the illness in man.

For infectious diseases, a valid animal model must meet certain criteria. The animal must be susceptible to the organism by at least one method of exposure, preferably the same portal of entry that occurs in man. Moreover, the microbe must survive, grow, and be readily recovered. Of course, administration of the organism must yield results that are reproducible. Then there are practical matters: the size of the animals, ease of handling, dietary requirements, housing facilities, cost, and so forth. A method of inoculation must be as simple as possible to minimize side effects. For research dealing with legionellosis, biocontainment is an additional concern.

A. Biocontainment

1. General Considerations

Griesemer and Manning[1] describe the precautions to minimize risks in an animal laboratory. One should purchase experimental animals that are known to be free of contagious diseases. They should be quarantined and conditioned before use. After quarantine the animals should be housed in isolation facilities where they are unlikely to become diseased. A program should be instituted for continous surveillance for contagious diseases. The major emphasis in containment should be at the source of contamination in the form of a primary barrier. The better the primary barrier, the less important the need for secondary barriers in the design and operation of the animal quarters. Finally and most important, a continuous educational program must be instituted to remind personnel of the principles of safety in the animal laboratory and the proper use of containment equipment and methods.

An animal laboratory facility should contain a number of biocontainment features. There should be a primary barrier at cage level and secondary barriers in room layout and airflow. In addition, ultraviolet or laminar flow airlocks and provision for steam or chemical sterilization of effluents are necessary to prevent contamination. Quarantine facilities, diagnostic laboratory support, and an incinerator are other essential features. A well-designed animal facility should also contain separate traffic flow for animals, people, and supplies, space and equipment for washing and sanitizing, service corridors, and an intercommunicating system.

The Office of Biosafety at the Centers for Disease Control has recommended minimal standards for handling laboratory animals infected or challenged with Legionella:[2]

1. It is recommended, but not required, that infected animals be maintained in animal rooms and an exhaust ventilation system that creates directional airflow into the animal room from adjacent clean areas and that the exhaust air be discharged to the outside and not recirculated to other rooms. Animal cages should be autoclaved before cleaning and recycling. Infected animal carcasses should be autoclaved in the immediate animal area; however, if this area has no autoclave, the material must be transported in a leakproof covered container to the autoclave or to a permitted pathological incinerator.
2. All procedures with a high potential for creating aerosols are conducted in a biological safety cabinet including inoculation and harvesting of embryonated hen's eggs with material known or suspected of containing Legionella. Ideally, necropsy of small laboratory animals is also conducted in a biological safety cabinet. If, however, the animal necropsy area is not provided with a biological safety cabinet, appropriate precautions should be taken by personnel to minimize exposure to infectious materials and contamination of the laboratory environment. These include: avoiding the creating

of aerosols and droplets, thorough decontamination of work surfaces, equipment and protective clothing following the necropsy, and incineration of the animal carcasses.

2. Selection of Animal Models

There are a series of factors that must be considered in deciding which animal species and strain should be used in biomedical research.[3] It is important to know the life-span of the species since the investigator may want to conduct a chronic or acute study. The researcher must determine the segment of the life-span best suited for the study such as newborn vs. aged animals. It is necessary to be aware of the anatomical characteristics, nutritional requirements and habits of the species. Polymorphism, genetic characteristics, and reproductive characteristics such as gestation, litter size, growth rate, and sex differences must also be taken into consideration. When selecting an animal species, the investigator must keep in mind various experimental parameters such as type of agent (drug, biohazard, or radiochemical), schedule of administration (frequency and quantity), and route of administration (aerosol, intubation, injection, or dietary). There are also practical matters such as available space and caging, supportive treatment (veterinary care and observations), and special environmental requirements for maintenance (temperature, humidity, light, noise).

3. Overview of Animal Models of Legionellosis

There are several models used for research into legionellosis that employ different animals and various methods of inoculation. The guinea pig is the most widely tested animal but some laboratories use rats, mice, rabbits and monkeys. Routes of inoculation include intraperitoneal, intratracheal, subcutaneous, respiratory aerosolization, and intranasal inhalation (nose drops). The organism has also been administered to animals in drinking water and directly intragastrically. The following sections summarize the major animal models including the description of experimental technique and a discussion on the merits and disadvantages of each one.

ANIMAL MODELS OF LEGIONELLOSIS

Animal	Type	Method of inoculation
Guinea pig	*Cavia porcellus* (Hartley strain)	Intraperitoneal Intranasal Intratracheal Aerosol Subcutaneous Gastric intubation Drinking water
Rat	Sprague-Dawley	Intratracheal Intraperitoneal
Mouse	ICR (Institute of Cancer Research	Intraperitoneal Subcutaneous
	AKR/J	Intraperitoneal
Rabbit	New Zeland White	Intravenous
Monkey	Cynolmolgus Rhesus	Aerosol

4. Susceptibility

The majority of animal models of legionellosis involve guinea pigs, but other animals are susceptible to infection. Patton[4] performed a study evaluating the relative susceptibility of *Legionella pneumophila* in different animals. Highly susceptible species were the Hartley strain of guinea pig, Sherman rat, and Mongolian gerbil. Moderate to low susceptibility occurred in the ICR (Institute for Cancer Research) mouse, Syrian hamster, and New Zealand white rabbit. The white leghorn chicken, coturnix quail, and courneaux pigeon were not susceptible. (All of the animals were inoculated intraperitoneally.) Collins[5] demonstrated a high prevalence of antibody to *Legionella pneumophila* in horses, suggesting that horses are infected with *Legionella* commonly. Other animals such as cattle, sheep, pigs, and goats exhibited antibodies to *Legionella*.

II. GUINEA PIGS

To date, the male adult guinea pig (Hartley strain) is the principle animal used by investigators to study the pathobiology of legionellosis. The animal is susceptible to *Legionella* via various routes of infection and allows for the survival, growth and recovery of the microbe. Infection with *Legionella* produces the characteristic clinical and pathological features of Legionnaires' disease although the degree of parallelism with human disease varies according to the method of inoculation . Guinea pigs are also relatively easy to handle. They are small enough for experimentation under a biocontainment hood and are reasonably prolific.

Berendt[6] ascribed several disadvantages to the guinea pig model. Blood is relatively difficult to obtain, and the technique of cardiac puncture is hazardous to the host. Moreover, bacteremic and leukocyte counts are not uniform in infected guinea pigs. Despite these drawbacks, the guinea pig model seems to be the one that is best worked out at this time.

A. Intraperitoneal

The first animal model described in the literature was interaperitoneal inoculation of guinea pigs with *Legionella pneumophila*. Chandler inoculated guinea pigs intraperitoneally[7]with 1 mℓ of 10% suspension of infected lung tissue prepared in sucrose-phosphate-glutamate buffer, pH 7.2[8] obtained from a fatal case of Legionnaires' disease.

The guinea pigs were sacrificed 3 to 7 days after inoculation and 10% spleen suspension in sucrose-phosphate-glutamate buffer from these animals were inoculated into yolk sacs of embryonated hens' eggs. The embryos died within 4 to 7 days and yolk sacs from the infected eggs contained 1×10^6 eggs (LD_{50}) of *Legionella pneumophila* per milliliter. Katz and Hashemi[9] have also studied this route of inoculation.

1. Pathology

The principle gross lesion is diffuse peritonitis of varying severity. Microscopically, the peritoneum is covered by a mixed inflammatory infiltrate of macrophages, neutrophils, fibrin, and cellular debris. Microscopic changes in the lungs involve all lobes and consist of focal interstitial infiltration of mixed inflammatory cells with variable necrosis. Alveolar spaces contiguous to areas of interstitial inflammation are filled with neutrophils, macrophages, cellular debris and little or no fibrin. In the spleen, multifocal nodular reticuloendothelial cell hyperplasia with central necrosis and infiltration of neutrophils are commonly observed. The livers of some, but not all, animals contained sharply delimited areas of necrosis involving any portion of the hepatic lobule. Necrotic foci were moderately infiltrated with neutrophils and macrophages. The subcapsular sinuses of lymph nodes were distended with intact and necrotic inflammatory cells. Foci of mixed inflammatory cells were infrequently found in the epicardium and myocardium.

2. Efficacy

Intraperitoneal administration of *Legionella* is easy because it requires only a single injection of inoculum; the organism can be placed directly into the peritoneal cavity and the dose can be readily controlled. By this route, a sufficient concentration of *Legionella* can cause progressive, ultimately fatal infection. This model is a valuable positive control in studies employing other routes of inoculation since it readily substantiates virulence of the microbe.

The major drawback to this model is its pathological discrepancies to human disease. Evidence indicates that *Legionella pneumophila* is transmitted to man from aqueous environmental sources by inhalation.[6,10-15] Therefore, the intraperitoneal route is somewhat irrelevant to legionellosis of man.

The clinical features of the infection produced by intraperitoneal inoculation of *Legionella* only in part parallel that seen in humans, but the histologic features of the pneumonitis in guinea pigs with Legionnaires' disease differs from the human pulmonary findings.[7] The animal lesions are more focal, the interstitium is prominently involved early on, and fibrin and necrosis of alveolar exudate, characteristic of the human lesion, are minimal in the guinea pig. Perhaps even more important, suppurative peritonitis, a typical feature of guinea pigs inoculated intraperitoneally with *Legionella* is not a usual manifestation of Legionnaires' disease in humans. Some claim that *Legionella* is less virulent by the intraperitoneal route than by other routes, particularly aerosol.[6]

In summary, the intraperitoneal model is valuable for recovery of *Legionella pneumophila* from infected organs and testing virulence of an inoculum. Although there are epidemiological and pathological differences between this model and human Legionnaires' disease, intraperitoneal inoculation of *Legionella pneumophila* will produce clinical illness and tissue injury that may be useful in clarifying pathogenetic mechanisms of legionellosis.

B. Intranasal

Guinea pigs can be infected intranasally with legionellosis as described by Katz and Hashemi.[9] The inoculum is *Legionella pneumophila* serogroup 1 grown on buffered charcoal yeast extract agar,[16] passed in embryonated hens' eggs at least one time and suspended in Bovarnicks sodium phosphoglutamate.[8] The LD_{50} of the inoculum administered is 5×10^6.

Inoculation is accomplished by nose dropper inserted deep into the nasal passages in unanesthetized animals. Each animal is kept separately under a biocontainiment hood for 10 min prior to inoculation and for 2 hr after inoculation. (Katz has also developed a chronic intranasal animal model in which guinea pigs were inoculated 8 times intranasally with *Legionella pneumophila* (1×10^9 organisms) over a 2 month period.[44])

1. Pathology

By light microscopy, lung tissue exhibits focal, acute, peribronchiolar pneumonia characterized by alveolar septal edema and congestion, fibrin within alveolar spaces, and intraalveolar and septal infiltration by neutrophils and macrophages. The bronchial and bronchiolar mucosa are intact, but bronchiolar lumens often contain an exudate of macrophages, neutrophils and fibrin, surrounded by zones of consolidation (Figures 1 and 2). There is a splenitis, most prominent in the red pulp, consisting of neutrophils and macrophages (Figure 2). Hepatic lesions include microabscesses, focal coagulative necrosis, fatty change, and edema of hepatocytes. Kidneys occasionally display focal acute tubular necrosis, focal proliferation of glomerular mesangial cells, and marked interstitial congestion. Lymphadenitis and suppurative pericarditis are infrequent lesions.

2. Efficacy

This animal model exposes the guinea pig to the organism via the respiratory route. The infection produced by this method of inoculation closely resembles the human form of legionellosis although the pathology is more systemic in guinea pigs than in humans. Features typical of human infection, i.e., terminal bronchiolitis and focal confluent alveolitis, are consistently produced by the intranasal model.

It is relatively easy to administer *Legionella* by this method. Animals that sneeze, however,

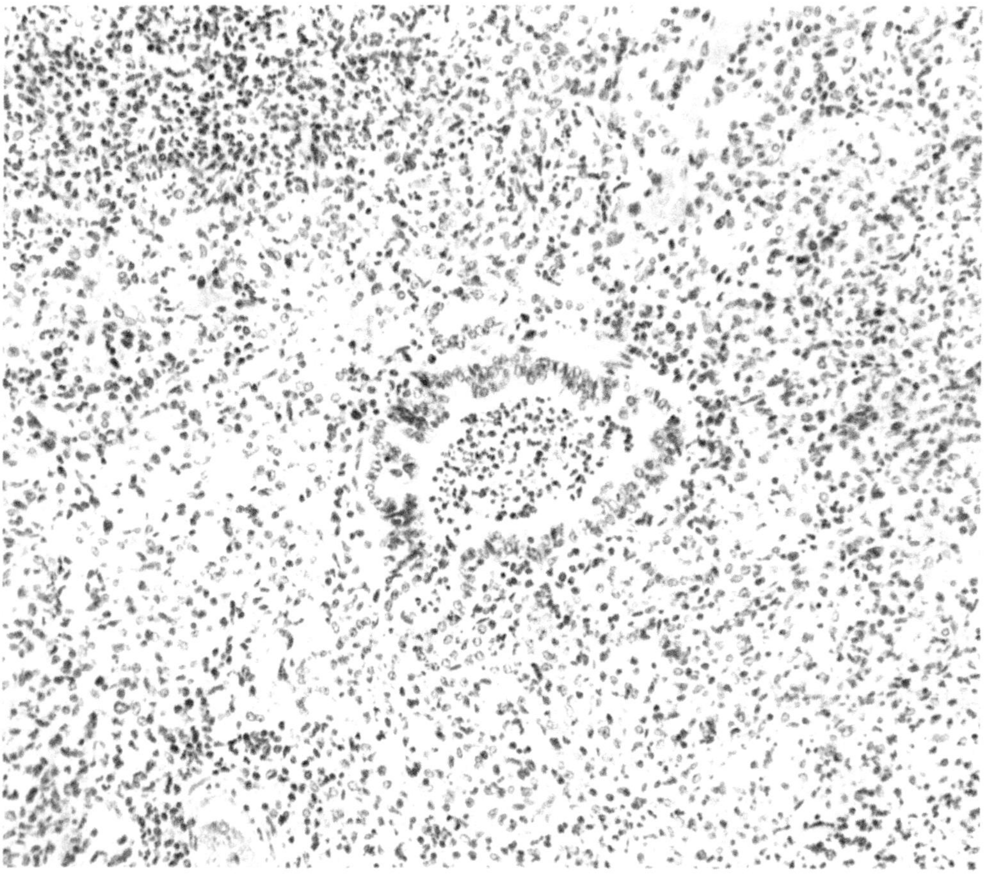

FIGURE 1. Peribronchial-bronchiolar consolidation features an exudate of neutrophils and macrophages. Bronchial mucosa is intact but plugged by inflammatory cells. (Hematoxylin and eosin, original magnification × 100.)

reduce the dose and are discarded from the study. It is also possible that animals might swallow some of the inoculum. Aerosols are not produced if the inoculum is placed directly into the upper respiratory tract. This is significant because it is believed that Legionnaires' disease is transmitted by inhalation of aerosolized *Legionella*[6, 10-15]

Despite these disadvantages, intranasal inoculation of *Legionella* is a useful animal model because it exposes guinea pigs to the organism via the natural route of infection and produces clinical and pathological findings somewhat comparable to human legionellosis.

C. Intratracheal

Winn has described the production of Legionnaires' pneumonia in guinea pigs inoculated intratracheally with *Legionella pneumophila*.[17] The inoculum was *Legionella pneumophila* serogroup 1, recovered from a patient with pneumonia. After passage of the organism twice on buffered charcoal yeast extract (BCYE) agar[16] the bacterium was inoculated into a biphasic medium that was prepared by overlaying BCYE agar with a semisynthetic broth medium for cultivation of *Legionella*[18] incubated at 35°C in room air and stirred using a heatless magnetic stirrer. Guinea pigs were inoculated intraperitoneally with approximately 10^5 colony forming units (CFU) of bacteria (LD_{50}) and sacrificed either when moribund or after fever for 2 days. The spleens were harvested and homogenized at a 10% suspension in phosphate-buffered saline. After three passage cycles through biphasic medium and guinea pigs, multiple aliquots of a pool of guinea pig spleens were prepared and frozen at -70°C.

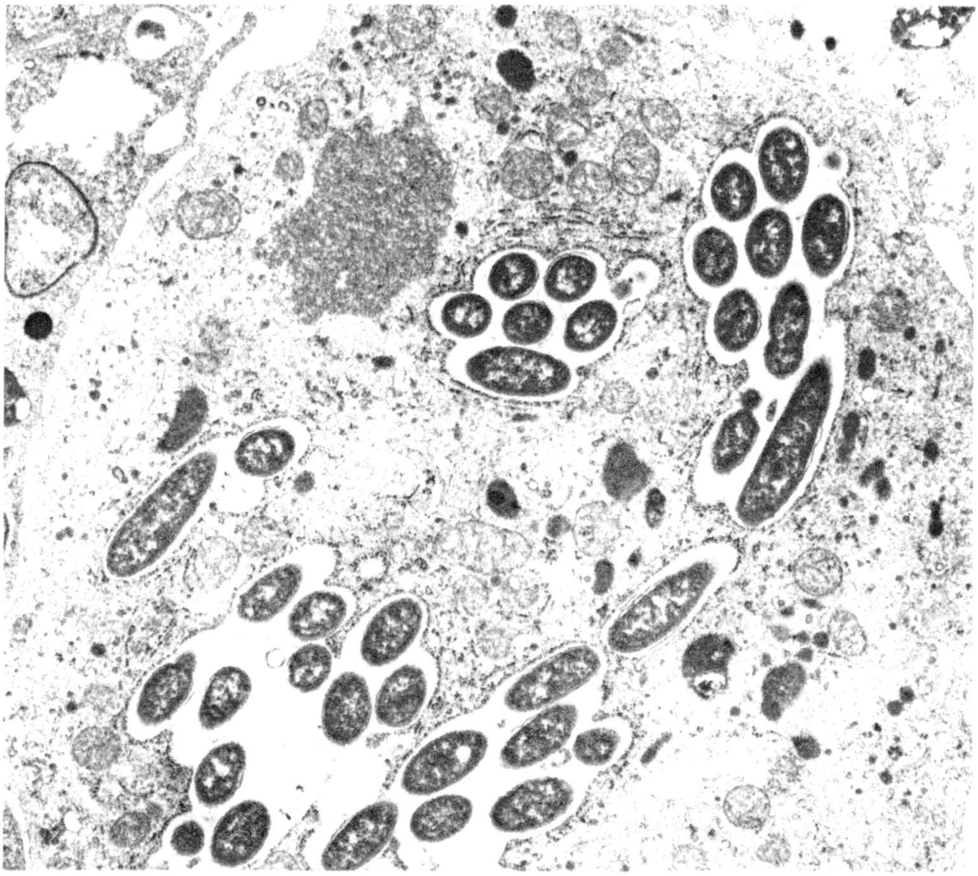

FIGURE 2. Sinusoids of the spleen of an infected guinea pig show an infiltrate of neutrophils and macrophages. (Hematoxylin and eosin, original magnification × 450.)

For inoculation, animals were anesthetized by intramuscular injection of a mixture of ketamine and acepromazine. The anterior tracheal wall was exposed by dissection, 0.5 mℓ of the bacterial suspension was injected into the lumen through a 25 gauge needle, and the incision was closed with steel clips.

1. Pathology

Macroscopic evidence of pneumonia was found in infected guinea pigs that died within the first 5 days. Multiple, tan nodular infiltrates, ranged in size from 1 to 5 mm, frequently were concentrated centrally, and were distributed along the bronchi and bronchioles. An acute inflammatory exudate was present in the distal air spaces of the guinea pigs examined microscopically. Extensive consolidation of pulmonary parenchyma was observed in most guinea pigs that died or were sacrificed within the first 5 days after inoculation. The exudate consisted of a mixture of polymorphonuclear neutrophils (PMNs) and macrophages with a small admixture of fibrin. Intraluminal aggregates of inflammatory cells and fibrin were observed in bronchioles and small bronchi, and scattered inflammatory cells were apposed to the luminal mucosa, but the inflammatory process infrequently involved the walls of the bronchial system. In some areas that contained exudate in the air spaces, the interstitium was widened and inflamed. Slight leukocytolysis of the inflammatory cells were present. The spleens consisted of focal accumulation of PMNs and macrophages in the red pulp. Less frequently small granulomas composed of clumps of macrophages alone were seen. Larger lesions included a central collar of macrophages. Coalescence of these granulomatous microabscesses obliterated the red pulp in some animals. Focal sinusoidal accumulations of macrophages were observed in the livers of infected guinea pigs. In some of these aggregates, an admixture of PMNs was present and there was slight leukocytosis. There were occasional small accumulations of inflammatory cells in the pericardium and myocardium. There was no pathology in the kidney.

2. Efficacy

This model is valuable because the guinea pig receives the inoculum via a respiratory route of infection. Like the intranasal model, the histologic pneumonia produced by intratracheal instillation somewhat resembles human Legionnaires' disease. Another advantage of this route is the ability to quantitate the inoculum accurately and assure that each animal has received the same dose.[17] Winn reports that the distribution of pneumonia in guinea pigs was sufficiently great in density and volume to insure widespread distribution of the bacteria. Finally there was evidence of infection in almost 100% of the guinea pigs inoculated intratracheally.

There are several disadvantages to this model. The peripheral administration of anesthetic agents depress respiratory function and potentially important early clearance mechanisms in the initial hours after infection.[17] Moreover, the surgical exposure of the trachea is a delicate procedure that requires a skilled technician who is qualified to do surgery in this anatomically complex part of the body. This procedure is also time consuming and exposes the guinea pig to infection by other agents. Finally, this procedure places the *Legionella* directly into the trachea and does not allow for the passage of *Legionella* through the nasal passage as occurs naturally in humans.

D. Inhalation of Aerosol

Inhalation of aerosolized *Legionella* is another common mode of infection.[6,10,19] There is a great deal of evidence suggesting an airborne transmission of *Legionella* in aerosols generated by air conditioning systems, cooling towers, evaporative condensers,[11,12,20,21] and other aqueous systems such as shower heads and faucets.[22-26] Thus, a model exposing pigs to *Legionella* by this method most closely parallels human infection.

Currently, there are several systems utilized for aerosolizing *Legionalla*. Davis reported a pathological pattern similar to intratracheal inoculation[17] in guinea pigs exposed to an aerosol by inhalation through a nose-only apparatus.[19]

Baskerville[10] exposed guinea pigs to aerosols of *Legionella pneumophila* in a Henderson apparatus. This machine allows control of relative humidity and introduces the aerosol directly into the respiratory tract. Colonies of *Legionella pneumophila* were washed off CYE plates and resuspended in distilled water. The guinea pigs were exposed for 5 min to bacterial aerosols generated with a three-jet collison spray at 65% relative humidity in the Henderson apparatus. The calculated inhaled doses for individual guinea pigs ranged from 2.4×10^3 viable bacteria to 4.9×10^5 viable bacteria (LD_{50}). At necropsy the lungs contained multiple foci of consolidation which became larger and coalesced to form extensive consolidation of all lung lobes. The histological lesions at 24 hr were numerous pneumonic foci with accumulation of neutrophils and macrophages in alveoli in all regions. In the later stages there was widespread edema, fibrin, exudation, hemorrhage, collapse, and infiltration of alveoli and interalveolar septa by equal numbers of neutrophils and macrophages.

Berendt experimentally infected guinea pigs with aerosols of *Legionella pneumophila* in an apparatus that resulted in whole body exposure instead of direct inoculation into the respiratory tract.[6] A collision atomizer and Henderson-type aerosol mixing and transit tube attached to a plastic exposure box was used for the exposure of guinea pigs to *Legionella pneumophila*. Aersol sampling was accomplished with all glass impingers containing 20 mℓ of tryptose saline solution (pH 7). The concentration of *Legionella pneumophila* in the sampler was estimated by spreading 0.1 mℓ of selected dilutions of impinger fluids on the surface of CYE agar plates and counting the colonies produced after suitable incubation. The median infectious dose was less than 129 organisms and the LD_{50} was 1.4×10^5 organisms. Berendt reported fever, weight loss, and seroconversion but lung lesions were not produced.

In general, animal models employing aerosols are valid because they are the best method of simulating airborne transmission of *Legionella*. Moreover, these animal models produce the typical features of Legionnaires' disease. The technique however, requires additional machinery and costly facilities of biocontainment.

E. Subcutaneous

Arko has developed a method for implanting stainless steel culture chambers into the subcutaneous tissue of the dorso-lumbar region of guinea pigs.[27] After 10 days, chambers in these animals are inoculated with *Legionella* using a syringe and needle with various dilutions of inoculum. After inoculation, fluid is removed from the chambers for bacteriological culture.

When laboratory-grown cells of Knoxville 1 and Pontiac 1 strains were inoculated into porous chambers implanted in normal guinea pigs, the animals developed nonlethal infections that persisted for median periods ranging from 9 to 12 days. The median infective dose ranged from 1×10^5 to 1×10^7 CFU. Slides of chamber fluid from Legionnaires' disease infected guinea pigs contained various numbers of intracellular and extracellular slender bacilli with tapered ends. The host inflammatory reaction included macrophages, neutrophils, lymphocytes, erythrocytes, fibrin, and other tissue debris.

As Arko points out, this animal model permits a wide range of research activities. The implanted chamber permits research on controlled numbers and types of microorganisms to be conducted with minimal discomfort or risk of contamination. Because the presence of the nonlethal *Legionella* infection is ascertained by culture of the chamber fluid, the problem with nonspecific or toxic substances that may kill chicken embryos or animals injected intraperiteneally is reduced. Within the chamber, infections may persist for several weeks, which permits study of the interaction of immunologic factors and microorganisms for an extended period of time. In addition, the tissue capsule surrounding implanted chamber appears to reduce host pressures on the inoculated organisms, thereby permitting the establishment of infections with laboratory-grown *Legionella* cells that may have lost virulence for other animal systems.

F. Drinking Water and Intragastric

Recently, Katz and Matus[28] developed an animal model in which a self-limited febrile illness is produced by oral inoculation of *Legionella pneumophila* into guinea pigs. It remains to be demonstrated whether the illness is a result of inhalation of an aerosol generated by drinking, aspiration of microbes during drinking, or infection of the gastrointestinal tract or some other mechanism. The research was preceded by a small-scale study by Conner and Gilbert.[29] The illness is accomplished by inoculating virulent *Legionella pneumophila* into the drinking water (1×10^3 to 1×10^8 organisms per milliliter) or inoculating the organism directly into the stomach (1×10^3 to 1×10^{10} organisms per milliliter) by means of autoclaved stainless-steel tubes. The inoculum was *Legionella pneumophila* serogroup 1, passed three times in embryonated hens eggs, streaked onto BCYE one time and suspended in Bovarnicks sodium phosphoglutamate.

Histologically, the lungs of these animals exhibited a mild, patchy, predominently interstitial and mainly macrophagic pneumonitis (Figure 4). Occasionally purulent bronchiolitis was observed. Splenitis featured congestion, sinus histiocytosis, scattered neutrophils, and rare microabscesses. Sections of the ileum occasionally showed focal aggregates of neutrophils and macrophages within the lumens and mucosa.

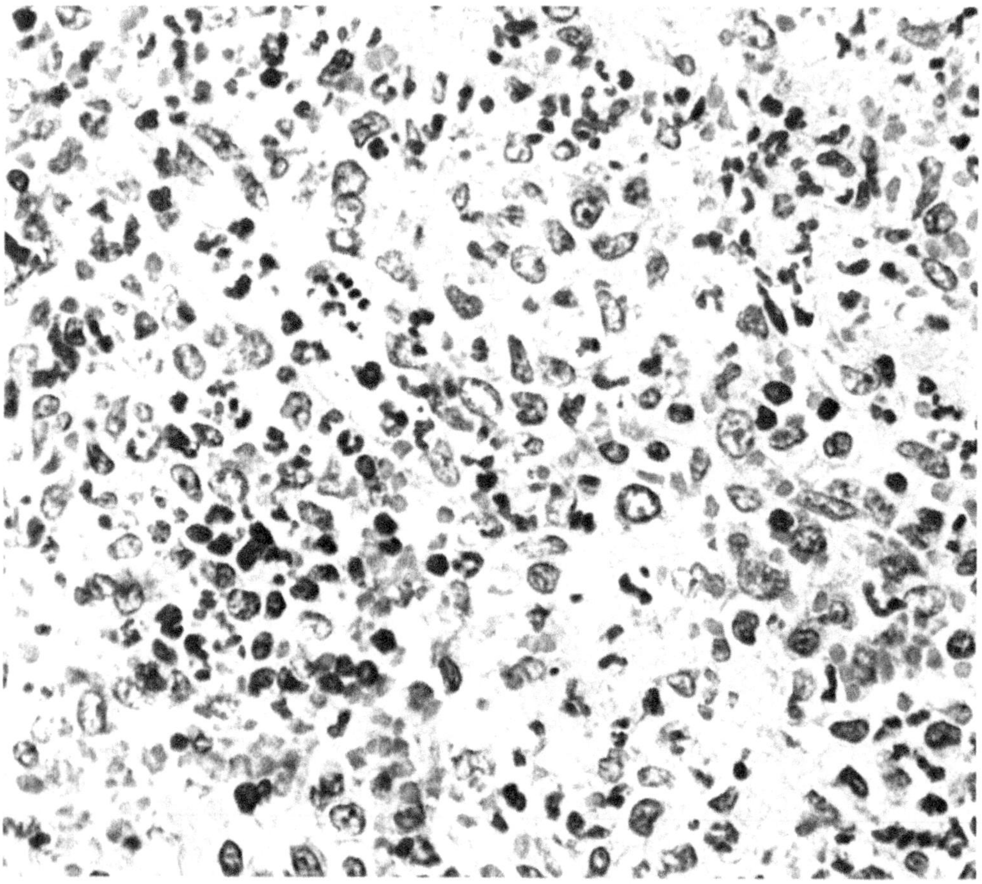

FIGURE 3. *Legionella pneumophila* is present within ribosome-studded membrane-bound vacuoles of a macrophage. These cytoplasmic organelles resemble rough endoplasmic reticulum. (Lead citrate, uranyl acetate, original magnification × 5000.)

Although the organism is administered by the gastrointestinal and not the respiratory route, this model can make valuable contributions to the understanding of Legionellosis. It might help to explain the prominent gastrointestinal symptoms often seen in Legionnaires' disease patients,[30-33] and might help to elucidate the epidemiology of legionellosis (Figure 3).

III. RATS

Lane-Petter and Pearson explain why rats and mice are ideal animals for laboratory research.[34] Their adaptability to the laboratory environment makes them an ideal choice for breeding and maintenance in captivity. They are omnivorous, robust, prolific, small, and, after a few generations of breeding in capitivity, tame. They are cheap, easy to breed and maintain and, in not being highly specialized, are useful in research relevant to human functions and diseases.

Winn[17] inoculated male, Sprague-Dawley rats intratracheally as described in the section on guinea pigs. The antibody response in rats and guinea pigs were comparable to microagglutination[35]and indirect immunofluorescence[36] techniques but the rats were less susceptible to lethal infections by *Legionella pneumophila*. Although the pulmonary pathology consisted of confluent macroscopic pneumonia, airspace infiltrate of neutrophils and macrophages, and leukocytosis, the pneumonia was less extensive in the rats. There was a

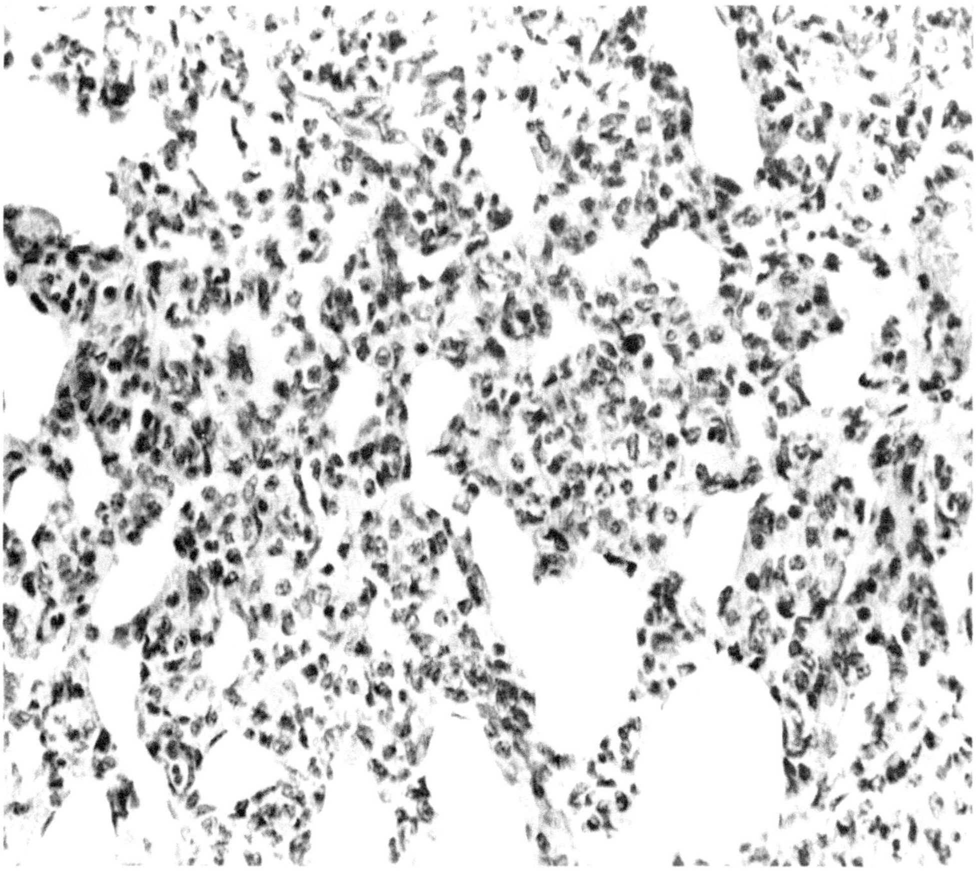

FIGURE 4 Alveolar septal thickening and an inflammatory infiltrate signify pulmonary interstitial pneumonitis. (Hematoxylin and eosin, original magnification × 250.)

paucity of extrapulmonary inflammation with one rat exhibiting small granulomas with central necrosis and microcalcification in the spleen. There were also 2 rats with sinusoidal macrophages in the liver.

Rolstad tested the susceptibility to infection with different strains of *Legionella pneumophila* in highly inbred rat strains Ho, DA and (Ho X DA)F_1 hybrids.[37] *Legionella pneumophila* serogroup 1 (Philadelphia strain) and *Legionella pneumophila* serogroup 1 (LD-8 strain) were harvested from 3 day CYE culture in physiological saline directly from CYE plates (the LD-8 strain was isolated from the Legionnaires' disease epidemic in Vasteraas, Sweden.) When about 9×10^8 bacteria or 6×10^8 colony-forming units were injected intraperitoneally, only rats infected with LD-8 strain showed signs of illness. The LD_{50} for this highly pathogenic strain was 3×10^8 CFU.

Macroscopic and microscopic examination of lungs, liver, spleen, gut, and kidneys of rats injected i.p. with Philadelphia 1 strain of *Legionella pneumophila* revealed no visible pathological changes except that some of the rats developed a marked splenomegaly. In contrast, several pathological changes could be detected histologically in rats injected with LD-8 strain of *Legionella pneumophila*. Scattered microabscesses were found in the liver lobules and in the adrenal cortex. The liver also showed moderate centroacinar vacuolization of the hepatocytes. The endocardium exhibited focal leukocytes infiltration indicative of endocarditis. The spleen white pulp was characterized by a pronounced lymphocyte depletion. The lungs did not show any distinct inflammation, but did show a focal congestion with interstitial edema. The intestines and the kidneys appeared normal.

In summary, the rat is a good alternative model for legionellosis because as Winn suggests, the rat may represent the model of a more resistant host in which the resistance to legionellosis in humans can be studied. The rat is also valuable for investigating the pathogenicity of different strains of *Legionella pneumophila* and possibly different strains of other species of *Legionella*. Finally, all of the previously mentioned factors for the popularity of rats as laboratory animals, make this a practical model.

IV. MICE

Mice have occasionally been used as animal models to study legionellosis.[4,27,38] Patton reported a moderate susceptibility in ICR mice after intraperitoneal challenge.[4] Arko[27] inoculated female ICR mice subcutaneously as described in the section on guinea pigs. When laboratory-grown cells of Knoxville 1 and Pontiac 1 strains were inoculated into porous chambers implanted in normal mice, the animals developed nonlethal infections that persisted for median periods ranging from 7 to 17 days. The median infectious dose was 1×10^7 CFU. As in the guinea pigs, slides of chamber fluid from Legionnaires' disease-infected mice contained various numbers of intracellular and extracellular slender bacilli with tapered ends. The most inflammatory reaction included macrohages, neutrophils, lymphocytes, erythrocytes, fibrin, and other tissue debris. Hedlund[38] showed that female AKR/J mice are susceptible to an intraperitoneal inoculation with virulent *Legionella pneumophila* and that 2×10^8 organisms cause death within 24 hr. The histopathologic findings included pyknotic and karyorrhetic cells in the spleen and thymus, neutrophilic infiltrates in the intestine, and foci of coagulative necrosis and vacuolated hepatocytes in the liver.

The mouse animal model has merit because *Legionella* infection can be produced in an animal that is practical and suitable for laboratory investigation. Also, the histological lesions somewhat resemble those seen in humans.

V. RABBITS

Baine reports male, New Zealand white rabbits being infected with *Legionella pneumophila* by intravenous administration of allantoic fluid from eggs infected with *Legionella pneumophila*.[39] Two mℓ of infected allantoic fluid containing 3×10^7 to 2×10^8 CFU of *Legionella pneumophila* was injected into a marginal ear vein in each test animal. The rabbits had a febrile illness and many died within 2 days of inoculation. At autopsy, the animals had enlarged spleens and scattered areas of pulmonary consolidation. Lungs and spleens from the test animals contained *Legionella pneumophila* detectable by direct immunofluorescence,[40] culture,[16] and urine from these animals was sometimes positive for the antigen of *Legionella pneumophila*, as determined by ELISA technique.[41] Interestingly, attempts to induce illness in rabbits by intranasal administration of *Legionella pneumophila* were less successful. Baine suggests that may be due to the fact that intranasal admistration of bacteria is more analogous to aspiration of pharyngeal contents than to inhalation of an aerosol, which is thought to be the principal way in which naturally occurring Legionnaires' disease is acquired.[42]

The rabbit animal model is worthy because rabbits are easily and cheaply obtainable. They are also good for dissection because of their size. Finally it is much easier to obtain blood, for serologic testing, from rabbits than from the animals discussed above. The analogy to human illness needs to be elucidated.

VI. MONKEYS

The cynolmolgus monkey[43] and Rhesus monkey[10] have been infected by inhalation of aerosol containing *Legionella pneumophila*. The preliminary studies indicated that cynolmolgus monkeys showed clinical signs of illness after the aerosol induced challenge. The Rhesus monkeys were exposed for 10 min to bacterial aerosols generated with a three-jet collison spray at 65% relative humidity in a Henderson apparatus.[10] The calculated inhaled dose for the monkeys was 6×10^6 viable organisms. They had slight fever and numerous scattered pneumonic foci in peripheral lung tissue. Affected alveoli and interalveolar septa contained large numbers of neutrophils and macrophages. A few associated respiratory bronchioles also contained exudate, but bronchi and bronchioles were normal.

Although monkeys and other primates are the most similar animals to man this animal model is not feasible because they are difficult to handle, have specialized dietary and environmental requirements, may carry infectious organisms pathogenic to man, and are extremely costly.

REFERENCES

1. **Griesemer, R. A. and Manning, J. S.,** Animal facilities, in *Biohazards in Biological Research,* Hellman, A., Oxman, M., and Pollack, R., Eds., Cold Spring Harbor Laboratory, Cold Spring Harbor, N.Y., 1973, 316.
2. **Jaugstetter, J. E.,** Recommendations for Biocontainment and Safe Handling of *Legionella pneumophila* and other *Legionella* Species, Memorandum, Centers for Disease Control, Atlanta, August 10, 1981.
3. **LeMunyan, C.,** Selection of the Appropriate Biological Model for Research in Environmental Health. Proc. 2nd Symp. Animal Models for Biomedical Research II, National Academy of Sciences, Washington, D.C., 1969, 27.
4. **Patton, C. M., Johnson, S. E., and Kaufmann, A. F.,** Susceptibility of laboratory rodents, birds, and rabbits to *Legionella pneumophila* (Legionnaires' disease bacterium), *Abstr. Annu. Meet. Am. Soc. Microbiol.,* B 79, 28, 1979.
5. **Collins, M. T., Cho, S. N., and Reif, J. S.,** Prevalence of antibodies to *Legionella pneumophila* in animal populations, *J. Clin. Microbiol.,* 15, 130, 1982.
6. **Berendt, R. F., Young, H. W., Allen, R. G., et al.,** Dose-response of guinea pigs experimentally infected with aerosols of *Legionella pneumophila, J. Infect. Dis.,* 141, 186, 1980.
7. **Chandler, F. W., McDade, J. E., Hicklin, M. D., et al.,** Pathologic findings in guinea pigs inoculated intraperitoneally with the Legionnaires' disease bacterium, *Ann. Intern. Med.,* 90, 671, 1979.
8. **Bovarnick, M. R., Miller, J. C., and Snyder, J. C.,** The influence of certain salts, amino acids, sugars and proteins on the stability of Rickettsiae, *J. Bacteriol.,* 59, 509, 1950.
9. **Katz, S. M. and Hashemi, S.,** Electron microscopic examination of the inflammatory response to *Legionella pneumophila* in guinea pigs, *Lab. Invest.,* 46, 24, 1982.
10. **Baskerville, A., Fitzgeorge, R. B., Broster, M., et al.,** Experimental transmission of Legionnaires' disease by exposure to aerosols of *Legionella pneumophila, Lancet,* 2, 1389, 1981.
11. **Cordes, L. G., Fraser, D. W., Skaliy, P., et al.,** Legionnaires' disease outbreak at an Atlanta, Georgia country club: evidence for spread from an evaporative condenser, *Am. J. Epidemiol.,* 111, 425, 1980.
12. **Dondero, T. J., Rendtorff, R. C., Mallison, G. F., et al.,** An outbreak of Legionnaires' disease associated with a contaminated air-conditioning cooling tower, *N. Engl. J. Med.,* 302, 365, 1980.
13. **Fisher-Hoch, S. P., Bartlett, C. L. R., Tobin, Jo'H., et al.,** Investigation and control of an outbreak of Legionnaires' disease in a district general hospital, *Lancet,* 1, 932, 1981.
14. **Fraser, D. W.,** Legionnaires' disease: four summers' harvest, *Am. J. Med.,* 68, 1, 1980.
15. **Politi, B. D., Fraser, D. W., Mallison, G. F., et al.,** A major focus of Legionnaires' disease in Bloomington, Indiana, *Ann. Intern. Med.,* 90, 587, 1979.
16. **Pasculle, A. W., Feeley, J. C., Gibson, R. J., et al.,** Pittsburgh pneumonia agent: direct isolation from human lung tissue, *J. Infect. Dis.,* 141, 727, 1980.
17. **Winn, W. C., Davis, G. S., Gump, D. W., et al.,** Legionnaires' pneumonia after intratracheal inoculation of guinea pigs and rats, *Lab. Invest.,* 47, 568, 1982.
18. **Pine, L., George, J. R., Reeves, M. W., et al.,** Development of a chemically defined liquid medium for growth of *Legionella pneumophila, J. Clin. Microbiol.,* 9, 615, 1979.
19. **Davis, G. S., Winn, W. C., Gump, D. W., et al.,** Legionnaires' pneumonia after aerosol exposure of guinea pigs and rats, *Am. Rev. Respir. Dis.,* in press, 1983.

20. **Kaufmann, A. E., McDade, J. E., Patton, C. M., et al.,** Pontiac fever: isolation of the etiologic agent (*Legionella pneumophila*) and demonstration of its mode of transmission, *Am. J. Epidemiol.*, 114, 337, 1981.
21. **Miller, R. P.,** Cooling towers and evaporative condensers, *Ann. Intern. Med.*, 90, 667, 1979.
22. **Cordes, L. G., Wiesenthal, A. M., Gorman, G. W., et al.,** Isolation of *Legionella pneumophila* from hospital shower heads, *Ann. Intern. Med.*, 94, 195, 1981.
23. **Dennis, P. J., Taylor, J. A., Fitzgeorge, R. B., et al.,** *Legionella pneumophila* in water plumbing systems, *Lancet*, 1, 949, 1982.
24. **Makela, T. M., Harders, S. J., Cavanagh, P., et al.,** Isolation of *Legionella pneumophila* (serogroup 1) from shower-water in Ballarat, *Med. J. Aust.*, 1, 293, 1981.
25. **Stout, J., Yu, V. L., Vickers, R. M., et al.,** Ubiquitousness of *Legionella pneumophila* in water supply of a hospital with endemic Legionnaires' disease, *N. Engl. J. Med.*, 306, 466, 1982.
26. **Tobin, J., Beare, J., Dunnill, M. S., et al.,** Legionnaires' disease in a transplant unit: isolation of the causative agent from shower baths, *Lancet*, 2, 118, 1980.
27. **Arko, R. J., Wong, K. H., and Feeley, J. C.,** Immunologic factors affecting the in-vivo and in-vitro survival of the Legionnaires' disease bacterium, *Ann. Intern. Med.*, 90, 680, 1979.
28. **Katz, S. M. and Matus, J. P.,** A Febrile Illness Produced in Guinea Pigs by Oral Inoculation of *Legionella pneumophila* (Abstr.), paper presented at the 2nd Int. Symp. *Legionella*, Atlanta, Ga., June, 1983.
29. **Conner, R. W. and Gilbert, D. N.,** *Legionella pneumophila* in drinking water of guinea pigs, *Ann. Intern. Med.*, 91, 323, 1979.
30. **Dournon, E., Bure, A., Kemeny, J. J., et al.,** *Legionella* peritonitis, *Lancet*, 1, 1363, 1982.
31. **Holt, P.,** Legionnaires' disease and abscess of appendix, *Br. Med. J.*, 282, 1035, 1981.
32. **Kirby, B. D., Snyder, K. M., Meyer, R. D., et al.,** Legionnaires' disease: clinical features of 24 cases, *Ann. Intern. Med.*, 89, 297, 1978.
33. **Swartz, M. N.,** Clinical Aspects of Legionnaires' disease, *Ann. Intern. Med.*, 90, 492, 1979.
34. **Lane-Petter, W. and Pearson, A.,** Utilization, *in The Laboratory Animal — Principles and Practice*, Academic Press, New York, 1971, 2.
35. **Farshy, C. E., Klein, G. C., and Feeley, J. C.,** Detection of antibodies to Legionnaires' disease organism by microagglutination and microenzyme-linked immunosorbent assay tests, *J. Clin. Microbiol.*, 7, 327, 1978.
36. **Wilkinson, H. W., Cruce, D. D., Fikes, B. J., et al.,** Indirect immunofluorescence test for Legionnaires' disease, in Legionnaires' the Disease, the Bacterium, and Methodology, Jones, G. L. and Hebert, G. A., Eds., HEW Publ. (CDC) 79-8375, Centers for Disease Control, Atlanta, 1979, 111.
37. **Rolstad, B. and Berdal, B. P.,** Immune defense against *Legionella pneumophila* in rats, *Infect. Immun.*, 32, 805, 1981.
38. **Hedlund, K. W., McGann, V. B., Copeland, D. S., et al.,** Immunologic protection against the Legionnaires' disease bacterium in the AKR/J mouse, *Ann. Intern. Med.*, 90, 676, 1979.
39. **Baine, W. B., Rasheed, J. K., Maca, H. W., et al.,** Hemolytic activity of plasma and urine from rabbits experimentally infected with *Legionella pneumophila*. *Rev. Infect. Dis.*, 1, 912, 1979.
40. **Cherry, W. B. and McKinney, R. M.,** Detection of Legionnaires' disease bacteria in clinical specimens by direct immunofluorescence, in Legionnaires' the Disease, the Bacterium, and Methodology, Jones, G. L. and Hebert, G. A., Eds., HEW Publ. No. (CDC) 79-8375, Centers for Disease Control, Atlanta, 1979, 91.
41. **Berdal, B. P., Farshy, C. E., and Feeley, J. C.,** Detection of *Legionella pneumophila* antigen in urine by enzyme-linked immunospecific assay, *J. Clin. Microbiol.*, 9, 575, 1979.
42. **Eickhoff, T. C.,** Epidemiology of Legionnaires' disease, *Ann. Intern. Med.*, 90, 499, 1979.
43. **Kishimoto, R. A., Kastello, M. D., White, J. D., et al.,** In vitro interaction between normal cynolmolgus monkey alveolar macrophages and Legionnaires' disease bacteria, *Infect. Immun.*, 25, 761, 1979.
44. **Katz, S. M.,** unpublished data.

Chapter 8

TISSUE CULTURE SYSTEMS

Charles E. Benson

TABLE OF CONTENTS

I. TISSUE CULTURE SYSTEM

In vitro studies of the pathogenic properties of the facultative intracellular bacterial pathogen *Legionella penumophila* may be approached in three different systems: continuous cell culture lines, diploid cell strains or macrophage cultures. The limitations and advantages of these systems will be discussed later in this chapter. An in vitro infection experiment is designed with the same consideration as an animal infection experiment. Decisions must be made regarding the type of cell infected, preparation and size of the infecting dose, length of exposure to the inoculum, susceptibility of the cells to infection, method(s) of data collection, and the length of the experiment. It is quite easy in this type of experiment to get so involved with the techniques that one loses sight of the purpose of the study. The general techniques of tissue culture maintenance may be reviewed in Merchant et al.[1] and Kruse and Patterson.[2] The objective of this chapter is to describe the various in vitro culture systems used as model systems to study the pathogenic properties of *Legionella* and the host cell response.

A. Continuous Cell Lines

Legionella has been grown in monolayers of several continuous cell lines such as HeLa cells (human cervical carcinoma, Figure 1), Hep 2 cells (human carcinoma of the larynx) and McCoy cells (mouse synovial cells).[3] The extent of the growth was not comparable to that obtained with bacteriological media; however, the microbes did increase 1.5 to 2.0 log within 96 hr and only Hep 2 cells supported an additional log increase when titered at 168 hr. These assays were performed by an immunofluorescent assay and do not distinguish the percent of viable bacteria. However, daily examination of the cells by light microscopy revealed changes in the cells not present in uninfected cells.

B. Diploid Cell Strains

The diploid cell line of MRC-5 cells (human fetal lung fibroblasts) supports the intracellular growth of *Legionella*.[3,4] These cells have a longer doubling time than the continuous cell lines and require more attention in their in vitro maintenance and transfer. However, the fibroblasts are more representative of the lung tissue (presumed primary target of *Legionella*) and are not specialized for phagocytosis. Such cells may be used to delineate the dynamics of the invasion (microbial and/or host mediated) and intracellular survival and growth of the microbe. Wong et al.[4] found large numbers of *Legionella* within MRC-5 cells and presented evidence that the intracellular generation time of *Legionella* was approximately two hours. Our studies[3] with the MRC-5 cells confirmed the intracellular growth of *Legionella*. The increase in the number of intracellular *Legionella* was not as dramatic as observed by Wong et al.,[4] but the distinct differences in the infection protocol may account for this discrepancy.

C. Conditions for In Vitro Infection

The proper coordination of the bacterial and eukaryotic cell cultures is critical for establishing an in vitro infection. The bacteria may lose specific pathogenic properties when subcultured for short periods on bacteriological media. Indeed, such is the situation with *Legionella*. As McDade and Shepard[5] indicated, the growth of *Legionella* on enriched Mueller-Hinton agar decreased the virulence of the organism for guinea pigs. The situation with in vitro infections may be different. Growth on FG or CYE agar[6] did not appear to affect the infection of the tissue culture systems although definitive infection dose studies were not performed. We routinely prepared an inoculum by transferring a single colony from CYE agar to 6 mℓ prewarmed FG broth. The inoculated broth was incubated at 37°C and growth was monitored by reading the turbidity on a Klett colorimeter (with No. 54

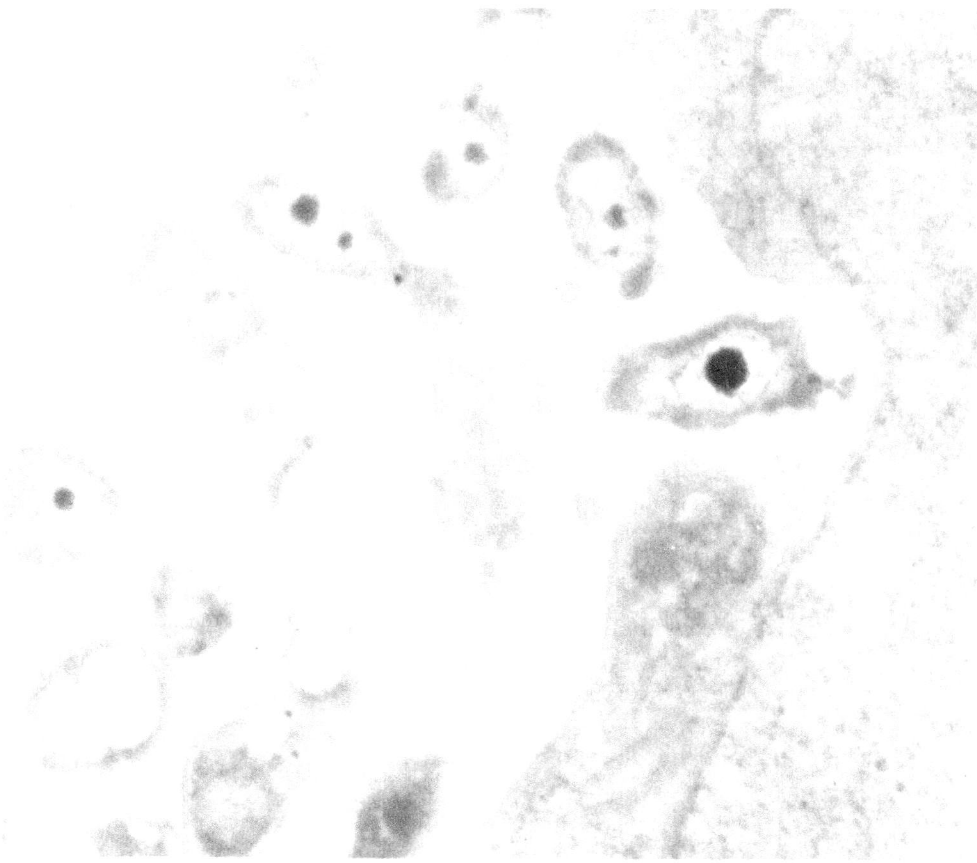

FIGURE 1. *Legionella pneumophila* surrounding and invading HeLa cells. (Magnification × 190,000.)

filter), or a Turner 330 spectrophotometer at 540 nm. A graph correlating the viable plate count (colony forming units) with the percent turbidity was constructed for each instrument and was used to estimate the increasing population. On the day of the experiment, the calculated CFU/mℓ of this culture was adjusted with fresh, prewarmed FG broth to 10^6 CFU/mℓ and 0.1 mℓ of the diluted bacterial culture added to the tissue culture tubes.

All tissue culture cells were grown as monolayers in 15 cm screw-capped tubes containing Eagles' minimal essential medium supplemented with 7.5% heat-inactivated fetal calf serum and 1% glutamine (hereafter called MEM). No antibiotics were added to the maintenance medium. The cells were routinely grown in 75 cm^2 tissue culture flasks (Corning) and subcultured to the screw-capped tubes and to confluency. The tissue culture medium was drained from the tube and the infection initiated by placing the 0.1 mℓ inoculum directly onto the monolayer. The tubes were incubated in cell culture tube racks at 37°C with 5% CO_2:95% air for 1 to 2.5 hr to permit attachment and infection. The volume of the inoculum was kept low to increase the incidence of contact of the microbe with the monolayers. After the attachment period, the broth was aspirated off and 2 mℓ of fresh prewarmed tissue culture medium added to each tube. The cultures were examined daily and the cells harvested at selected intervals.

We were concerned that the tissue culture media may support the extracellular growth of *Legionella*. Such growth could (1) lead to change of the constituents and/or pH of the tissue culture medium rendering it unfavorable for the eukaryotic cells, (2) increase the infecting dose, (3) continue the infecting process for an extended period, or (4) have no effect. Fresh and "conditioned" tissue culture medium was incubated with 10^6 CFU and incubated for

168 hr. The viable count fell to 0 after 24 hr although organisms (nonviable) could be detected by an immunofluorescent assay. Thus, *Legionella* were unable to replicate in fresh tissue culture or "conditioned" medium in the absence of viable eukaryotic cells. The conditioned medium was the solution harvested from a 4-day-old uninfected MRC-5 culture. The medium was plated to assure sterility. This control culture was necessary to alleviate concern about the possibility that metabolites and cell components released by eukaryotic cells could support the growth of the bacterium. Therefore, any increase in the number of viable microbes in the culture medium was the result of intracellular growth and release into the external milieu.

The infection protocol of Wong et al.[4] provided for a longer period of bacterium:host cell contact. He maintained the bacteria on MRC cells and prepared an inoculum by disrupting the infected cells with a tissue grinder and plating aliquots onto CYE agar. Following 48 hr incubation, the bacteria were suspended in Eagles' minimal essential medium. Leighton tube cultures (16 × 100 mm) were infected with approximately 10^5 to 10^6 bacteria contained in 1 mℓ MEM. The infection mixture was incubated at 37°C for 13 to 16 hr. At this time, the monolayer was washed with sterile Hanks balanced salt solution (HBSS) and fresh MEM containing 100 u of penicillin G sodium and 45 μg of streptomycin placed over the monolayer. These antibiotics were shown by the authors to kill all extracellular microbes within 3 hr. Similar data was obtained with gentamicin.[3] The use of antibiotics in both instances assured the sterility of the external environment and prevented reinfection of the monolayers. Changes in the integrity of the monolayer, pH, or turbidity of the medium could be attributed only to the intracellular growth of the microbe.

D. Monitoring Intracellular Growth of *Legionella*

The easiest method of noting effects of the infection is a change in the turbidity of the tissue culture medium and by microscopic inspection of the infected cells for any cytopathic effect (CPE). Both Wong et al.[4] and Daisy et al.[3] described progressive CPE of *Legionella* infected cells which begin with a granular appearance in the cytoplasm followed by disorganization of the monolayer within 24 to 72 hr following infection. Both research groups found the *Legionella* within membrane vacuoles in EM studies. Wong et al.[4] quantitated the microbes by scraping and disrupting the infected cells with a tissue grinder. The tissue extract was plated on CYE agar. We simply scraped the cells into 1 mℓ sterile distilled water with a blunt glass rod.[3] Lysis was complete within 3 to 5 min. The lysate was serially diluted and plated on CYE. Purity of the released bacteria was determined by immunofluorescent assay. The plate counts provide a relatively good estimation of the number of viable organisms and is generally lower than the total number determined by immunofluorescent techniques.[3] The methods summarized here document (1) the presence of cytopathology, (2) the intracellular existence of the pathogen, (3) the presence of the organisms within membrane-bound vesicles within the cytoplasm, (4) the intracellular replication, and (5) the homogenity of the organism from cell lysates. Contamination of the cell cultures or the inoculum will occur and must be carefully monitored by growth (or absence of growth) on solid media. The use of antibiotics in the cell culture media provides no absolute assurance against contamination nor does antibiotic use contribute to the experimental design.

E. Application

The use of tissue culture systems to evaluate host:parasite interaction affords the investigator the opportunity to manipulate the experimental system in a fashion not available in animal models. The appropriate type of cell for the work is critical. In the instance of *Legionella*, the MRC-5 cells are more appropriate than the continuous cell lines because MRC-5 cells are representative of the normal tissue parasitized by *Legionella*.

Through use of short adsorption:infection period, different inoculum concentrations and addition of antibiotics to the tissue culture medium to kill unadsorbed microbes, the investigator may achieve a limited or extensive infection of an appropriate monolayer. Multi-specimens may be obtained with relative ease, facilitating biochemical studies of the pathologic situation. The situation can be considerably more controlled than intact animal studies with less hazard to laboratory personnel. Data derived from tissue culture studies must be carefully constructed and interpreted with the awareness that validity of all conclusions must fit clinical observations and animal experimentation. Nevertheless, tissue culture infection studies, when properly performed, should reduce the number of costly animal experiments.

Our own application of these techniques involved an adaptation to generate antibiograms. We noted that the in vitro antimicrobial susceptibility correlated poorly with the in vivo clinical results and proposed that testing antimicrobial susceptibility of *L. pneumophila* in an intracellular environment might be better correlated with the clinical response. The study indicated no significant difference in the antibiogram data.[7] We concluded that a more appropriate system might have been monocyte cultures since these cells are the primary target cell during infection.[8]

Rinaldi et al.[9] have utilized animal cell cultures to investigate the pathogenic and cytotoxic properties of the Pittsburgh pneumonia agent (PPA), which resembles *Legionella*. They observed that the acid-fast characteristic of these anaerobes was associated only with those organisms grown in animal or cell culture systems. PPA grown on artificial media were not acid-fast. This staining characteristic may correlate with a specific virulent factor required for growth in host cells. The finding is important and indicates the potential of in vitro model systems for studies of pathogenic mechanisms.

II. MACROPHAGE CULTURES

A natural evolution of tissue culture studies of *Legionella pneumophila* has been the move to in vitro monocyte experiments. The monocytes may be isolated from a variety of sites (lungs, peripheral blood, peritoneum) and animals. The exciting potential of monocyte studies in studies of *Legionella* pathogenesis is being explored by only a few investigators. These approaches will be discussed. In each instance, the animal source is not sacrificed during the sampling period, an important consideration for further studies.

A. Isolation Techniques

1. Peripheral Blood Monocytes

Horwitz and Silverstein[10] diluted heparinized blood with 0.9% saline and centrifuged at 800 g for 30 min at 24°C over a Ficoll-sodium diatrizoate (Ficoll-Paque, Pharmacia). The layer containing the monocyte fraction was collected, diluted with an equal volume of RPMI-1640 containing glutamine (Gibco Laboratories) and the cells collected at 400 g for 10 min at 4°C. Several washes at 115 g (10 min, 4°C) were required to remove residual platelets. The cells were suspended in RPMI and counted. Horwitz reported greater than 99% viability (by trypan blue exclusion) in a mixed population of polymorphonuclear leukocytes, monocytes, and lymphocytes. The cells were placed in RPMI containing 15% autologous serum and incubated in 35 mm plastic petri dishes for 1 hr at 37°C in 5% CO_2:95% air. The plates were then washed with warm (37°C) RPMI and the adherent cells determined by counting a calibrated area. The average cell population was 2.2×10^6 cells per plate. No antibiotics were present in any medium.

2. Peritoneal Macrophages

Kishimoto et al.[11] obtained peritoneal macrophages from outbred guinea pigs (400 to 500 g) by the interperitoneal injection of 20 mℓ of 1.5% sodium caseinate (BBL). The peritoneal exudate was collected four days later and cultured on petri dishes or in Leighton tubes at 37°C for two hours. The cultures were washed twice with Hanks balanced salt solution

(HBSS) and fresh Earles' 199 medium containing 10% fetal calf serum added to the cell culture. The cell yield was not cited.

3. Alveolar Macrophages

Kishimoto et al.[12] utilized cynomolgus monkeys (2.0 to 3.5 kg, both sexes) as a source of alveolar macrophages. The chemically restrained animals were anesthetized with Halothane prior to anesthetizing the pharynx, larynx, and trachea with 2% lidocaine spray.[13] A cuffed endotracheal tube was positioned into the trachea and stabilized by inflating the cuff. The left lung was lavaged with 0.9% sterile saline (warmed to 37°C) passed from a sterile 50 mℓ syringe through the endotracheal tube. The wash was collected by withdrawal into the syringe and this wash cycle repeated with 5 more volumes of warm saline. The cells were collected by centrifugation at 230 g for 10 min and re-suspended in 10 mℓ Earles' 199 medium containing 10% homologous serum, penicillin (100 units/mℓ) and streptomycin (100 μg/mℓ). The mean total number of leukocytes was 2.87×10^6 per lavage and the macrophage population represented approximately 88 to 91% of this total. The cells were placed into 60 mm plastic dishes, incubated for 3 hr at 37°C at 5% CO_2, then washed twice with HBSS.

Dyer and Leid[14] have utilized a similar technique to harvest high numbers of alveolar macrophages from equine species. Ponies were lightly anesthetized and placed in right or left lateral recumbency. A double lumen tube was constructed using an equine nasogastric tube with an inner polyethylene tube. It was passed into the lung and sterile 0.85% saline containg penicillin (1 mg/mℓ) and dihydrostreptomycin (1 mg/mℓ) was used as the lavage fluid. Approximately 100 mℓ of fluid was introduced into the lung by gravity through the nasogastric tube, followed by a gentle aspiration of the fluid into a syringe fitted on the inner polyethylene tubing. The collected fluid was transferred to a centrifuge tube via a three-way stopcock. The lavage procedure was repeated until 300 mℓ of cell suspension had been collected. The cells were pelleted by centrifugation (400 g, 20 min, room temperature), washed three times in RPMI containing 10% fetal calf sera, penicillin (10 μg/mℓ) and streptomycin (100 μg/mℓ). The macrophages were further purified by plating into glass or plastic plates as in the above procedures. An average lavage yield was 12.4×10^7 purified macrophages. Dyer[21] had the same experience with lavages of bovine species. The animal is unharmed by this lavage procedure. We have further altered the procedure by placing 8×10^5 macrophages into each well of a 24-well tissue culture dish and proceeding with the 1 hr incubation and washes as above.[22] A variation on this procedure has been the addition of sterile 12 mm diameter coverslips to each well prior to the introduction of the cell suspension. The advantage to this modification is the ability to remove one or more coverslips at any time during an experiment for viability counts, lysis of the monolayer for quantitation of intracellular bacteria, fixation and staining or monitoring the levels of enzymes of the macrophage. Suspension cultures of bovine macrophages are also easily established but are more cumbersome and viability of the cells decreases at a more rapid rate than observed in attached cell cultures. We routinely observed over 90% viability after 3 to 4 days in culture.[22]

B. Conditions for In Vitro Infection

The recent papers of Kishimoto et al.[12] and Horwitz[10,15,16] are in accord concerning the preparation of the *Legionella* for the infection studies by growing the the pathogen in embryonated chick eggs. Kishimoto et al.[12] passed the organism through embryonated eggs three times and then grew it on enriched Mueller-Hinton prior to the infection studies. Horwitz et al.[10] inoculates the 6-day-old embryonated eggs with a 0.2% homogenate of a spleen from a guinea pig which had died following an interperitoneal injection of *Legionella*. Infected yolk sacs from embryos dying 4 to 8 days after infection were harvested, homogenized in 10 mℓ egg yolk buffer (EYB: 12 volumes 0.15 *M* KH_2PO_4, 38 volumes 0.15 *M* Na_2HPO_4, 50 volumes 0.85 NaCl, pH 7.2) and flash frozen in 1 mℓ aliquots prior to storage at −70°C. The bacteria were partially purified prior to use in experiments by diluting a quick-thawed aliquot in 10 mℓ EYB followed by centrifugation at 45 g for 10 min at 4°C. The supernatant was transferred to another tube and the bacteria pelleted at 1800 g for 20 min at 4°C. The bacterial pellet was twice washed with RPMI under the same conditions. The organisms utilized by Horwitz were never placed on artificial media prior to the experimental infections, thus decreasing the possibility of selecting for avirulent cells within the inoculum.

The EM studies of Kishimoto[11,12] involved short term (2 hour) incubations of the pathogen with the macrophages. The infecting dose was 100 organisms per macrophage. Specimens were then processed for transmission or scanning electron microscopy.

In the studies of Horwitz et al.,[10,15,16] the conditions for inoculation of the macrophage varied with each experimental protocol. In no instance was an overwhelming inoculation used. This is a particularly critical point since the experiments were often four to five days in duration.

C. Monitoring Cell:Bacterium Interaction

The evaluation of the host:cell:bacterium interaction has been rather unimaginative. Kishimoto et al.[11,12] concentrated their efforts toward descriptive analysis of electron photomicrographs while Horwitz et al.[10,15,16] have utilized bacteria survival and/or growth as an indicator. The macrophages were disrupted by a brief sonication burst which proved sufficient to lyse the cells and not damage the bacteria. Serial dilution of the sonicates were plated onto CYE agar. Immunofluorescent staining[6] of bacteria and infected macrophages were also used as evaluative tools.

D. Application

1. Pathogenic Mechanism

Virulent strains of *Legionella* survive and proliferate with the macrophages regardless of the source (animal species or resident habitat of the macrophage). Avirulent strains did not survive within the phagosome. These results have been documented by the electron microscopic studies[11,12] and plate counts.[10,15,16] The presence of specific anti-*Legionella* immunoglobulins did enhance phagocytosis; however, there was no evidence that the intracellular proliferation of the organism was diminished.

2. Host Response

Macrophages derived from guinea pigs previously inoculated with avirulent strains of *Legionella* were unable to retard the intracellular activity of the virulent strains.[12] Horwitz et al.[16] manipulated cultured macrophages to evaluate host response. Macrophage cultures were activated by incubation with concanavalin A (Con A) and lymphocytes (the nonadherent cell population). The activated macrophages phagocytized the organism and retarded the intracellular multiplication of the ingested pathogen. Cell-mediated immunity is suggested to play a major role in the host defense against the Legionnaire bacillus. The mechanism of this activated response remains to be elucidated. The combined use of animals and in vitro cultures provides the essential tools for continued studies.

III. LIMITATIONS AND ADVANTAGES OF THE IN VITRO SYSTEM

The choice of the ideal model system becomes clear as one moves from established cell cultures to the macrophage system. The macrophage was defined as the apparent target cell[8] and Horwitz has presented an exciting series of experiments. There are still many questions to answer about the activity of this cell during the infection process. However, *Legionella* invades the lungs and the activity of alveolar macrophages may well respond in a different fashion. The data of Kishimoto[11] may indeed suggest some differences. The convenience of a cell culture model permits multiple specimen handling (i.e., duplicate or triplicate samples per sampling time) with the generation of more reliable and consistent data. Experiments may be repeated at a faster rate dependent only on the rate of cell culture maintenance or isolation and number of cells required for the experiments. Further justification for in vitro studies is the relatively low cost per experiment compared to the cost, and problems, of maintaining a proper animal facility.

The environment of the cultured cells is easier to maintain (e.g., CO_2 environment, temperature, humidity, absence of antibiotics, addition of specific components to the infection) compared to all the micro and molecular interactions that occur with animals. This is not to imply that cell cultures are the best approach to study bacterial pathogenesis. Much like animals, the cultures often may become contaminated (e.g., sick), or die unexpectedly at the most inopportune moment or even appear to refuse to grow. The in vitro model system provides a mechanism to perform a series of critical experiments to develop a refined protocol for the final definitive studies within animals. Only in the animal, will the impact of various metabolites, hormones, activators, activators, humoral and cellular mechanisms, and normal microbial flora upon the host response to an assaulting pathogen be observed. Consider this: the epidemiological data of Collins et al.[17] implies the equine species may be appropriate although they have been unable to establish an infection in this animal species by intravenous or aerosal routes.[18] However, these animals were in possession of functional humoral and cellular immune systems. The data of Horwitz[16] suggests this to be adequate to handle a *Legionella* infection. Therefore, the immune capacity of the animal must be temporarily compromised to establish an infection by the natural route. Clearly there are many possibilities for further thought and experimentation.

IV. ADDENDUM: APPLICATION OF TISSUE CULTURE SYSTEMS

The use of cell culture systems to evaluate the events of the host:pathogen interaction continues to provide much insight on the pathological changes of *Legionella*-infected cells. Horwitz[19] has described the formation of an unusual vacuole within the first 4 to 8 hr after phagocytosis of the Legionellae. Cultured human peripheral blood monocytes were incubated with a high ratio of bacteria (approximately 5000 bacteria per monocyte) to achieve sufficient numbers of intracellular bacteria to permit EM studies of the initial host:parasite interaction. Fifteen minutes postinfection, greater than 95% of the phagocytic vacuoles were surrounded by smooth vesicles. Mitochondia were found around some vacuoles and the frequency of this observation increased with longer incubation. By 1 hr post-infection, many phagosomes were near an invagination of the nuclear envelope. At 4 and 8 hr postinfection, ribosomes were found lining the vacuoles. Only smooth phagosomes were observed when formalin-killed bacteria were used instead of viable *Legionella*. Therefore, the association of ribosomes with the phagosome is unique to the vacuoles containing viable microbes. Similar short-term studies have not been possible with the infected animal model. However, phagosome-associated ribosomes have been observed in EM studies (48 hr post-infection) of guinea pig alveolar macrophages.[20] This observation confirms the validity of the in vitro studies. The interesting movement of the phagosome and the dynamics of the initial phases of the pathogenic assault were observed only in the in vitro work. The biochemistry of this interaction should be delineated so the mechanism of the pathologic lesions can be understood.

REFERENCES

1. **Merchant, D. J., Kahn, R. H., and Murphy, W. H.,** Handbook of Cell and Organ Culture, *2nd ed., Burgess Publishing, Minneapolis, 1964.*
2. **Kruse, P. F., Jr. and Patterson, K. K., Jr., Eds.,** Tissue Culture: Methods and Applications, *Academic Press, New York, 1973.*
3. **Diasy, J. A., Benson, C. E., McKitrick, J., and Friedman, H. M.,** Intracellular replication of *Legionella pneumophila, J. Infect. Dis.,* 143(3), 460, 1981.
4. **Wong, M. C., Ewing, E. P., Jr., Callaway, C. S., and Peacock, W. L., Jr.,** Intracellular multiplication of *Legionella pneumophila* in cultured human embryonic lung fibroblasts, *Infect. Immun.,* 28(3), 1014, 1980.
5. **McDade, J. E. and Shepard, C. C.,** Virulent to avirulent conversion of Legionnaires' disease bacterium (*Legionella pneumophila*) its effect on isolation techniques, *J. Infect. Dis.,* 139, 707, 1979.
6. **Jones, G. L. and Hebert, G. A., Eds.,** *"Legionnaires,"* the Disease, the Bacterium and Methodology, Centers for Disease Control, Atlanta, Ga., 1979.
7. **Bacheson, M. A., Friedman, H. M., and Benson, C. E.,** Antimicrobial susceptibility of intracellular *Legionella pneumophila, Antimicrob. Agents Chemother.,* 20(5), 691, 1981.
8. **Chandler, F. W., Hicklin, M. D., and Blackman, J. G.,** Demonstration of the agent of Legionnaires' disease in tissue, *N. Engl. J. Med.,* 297, 1218, 1977.
9. **Rinaldo, C. R., Jr., Pasculle, A. W., Myerowitz, R. L., Gress, F. M., and Dowling, J. N.,** Growth of the Pittsburgh pneumonia agent in animal cell cultures, *Infect. Immun.,* 33(3), 939, 1981.
10. **Horwitz, M. A. and Silverstein, S. C.,** Legionnaires' disease bacterium (*Legionella pneumophila*) multiplies intracellularly in human monocytes, *J. Clin. Invest.,* 66(3), 441, 1980.
11. **Kishimoto, R. A., White, J. D., Shirey, F. G., McGann, V. G., Berendt, R. F., Larson, E. W., and Hedlund, K. W.,** *In vitro* responses of guinea pig peritoneal macrophages to *Legionella pneumophila, Infect. Immun.,* 31(3), 1209, 1981.
12. **Kishimoto, R. A., Kastello, M. D., White, J. D. Shirey, F. G., McGann, V. G., Larson, E. W., and Hedlund, K. W.,** *In vitro* interaction between normal cynomolgus monkey alveolar macrophages and Legionnaires' disease bacteria, *Infect. Immun.,* 25(2), 761, 1979.
13. **Kastello, M. D., Emmert, A. D., Denson, R. F., and Kishimoto, R. A.,** Recovery of alveolar macrophages from rhesus and cynomolgus macaques by long lavage, *Am. J. Vet. Res.,* 40, 271, 1979.
14. **Dyer, R. M., Ligget, D., and Leid, R. W.,** Isolation and partial characterization of equine alveolar macrophages, *Am. J. Vet. Res.,* 44(12), 2370, 1983.
15. **Horwitz, M. A. and Silverstein, S. C.,** Interaction of the Legionnaires' disease bacterium (*Legionella pneumophila*) with human phagocytes. II. Antibody promotes binding of *L. pneumophila* to monocytes, but does not inhibit intracellular multiplication, *J. Exp. Med.,* 153(2), 398, 1981.
16. **Horwitz, M. A. and Silverstein, S. C.,** Activated human monocytes inhibit the intracellular multiplication of Legionnaires' disease bacteria, *J. Exp. Med.,* 154(5), 1618, 1981.
17. **Collins, M. T., Cho, S. N., and Reif, J. S.,** Prevalence of antibodies to *Legionella pneumophila* in animal populations, *J. Clin. Microbiol.,* 15(1) 130, 1982.
18. **Cho, S. N., Collins, M. T., Reif, J. S., and McChesney, A. E.,** Experimental infections of horses with *Legionella pneumophila, Am. J. Vet. Res.,* 44(4), 662, 1983.
19. **Horwitz, M. A.,** Formation of a novel phagosome by the Legionnaires' disease bacterium (*Legionella pneumophila* in human monocytes, *J. Exp. Med.,* 158, 1319, 1983.
20. **Baskerville, A., Dowsett, A. B., Fitzgeorge, R. B., Hambleton, R., and Broster, M.,** Ultrastructure of pulmonary alveoli and macrophages in experimental Legionnaires' disease, *J. Pathol.,* 140, 77, 1983.
21. **Dyer, R. M.,** personal communication.
22. **Dyer, R. M. and Benson, C. E.,** unpublished.

Chapter 9

HOST RESISTANCE TO *LEGIONELLA PNEUMOPHILA*: INTERACTIONS OF *L. PNEUMOPHILA* WITH LEUKOCYTES

Marcus A. Horwitz

TABLE OF CONTENTS

I. INTRODUCTION

Legionella pneumophila has provided its share of surprises to the scientific and medical communities not the least of which is its capacity to multiply intracellularly in human mononuclear phagocytes. This capability is central to the pathogenesis of Legionnaires' disease and confronts the host immune defense system with a special challenge. This chapter focuses on the intracellular site of *L. pneumophila* replication and on the interactions in vitro between *L. pneumophila* and cellular and humoral components of the host immune defense system.

II. INTRACELLULAR MULTIPLICATION OF *L. PNEUMOPHILA*

L. pneumophila has been reported to multiply in vitro in the presence of human peripheral blood monocytes (Figure 1), human alveolar macrophages,[2] monkey alveolar macrophages,[3,4] guinea pig macrophages,[5] and several cell lines.[6,7] *L. pneumophila* does not multiply in association with human lymphocytes[1] or polymorphonuclear leukocytes.[8]

Several lines of evidence indicate that *L. pneumophila* multiply intracellularly in monocytes and not simply in association with them.[1] First, in the absence of leukocytes, *L. pneumophila* does not grow in tissue culture medium with or without serum even if the medium is conditioned for 4 days by viable monocytes. Second, *L. pneumophila* does not multiply in sonicated mononuclear cells. Moreover, *L. pneumophila* requires intact monocytes for multiplication at every phase of its growth. Lysis of infected monocytes at various times during logarithmic growth of *L. pneumophila* is followed by cessation of bacterial multiplication. Growth resumes when intact mononuclear cells are added back to the cultures. Third, in experiments utilizing parabiotic chambers, *L. pneumophila* multiplies only when in direct contact with monocytes. In these experiments, the two compartments of each parabiotic chamber were separated by a filter that contained pores large enough to allow ready passage of molecules but too small to allow passage of bacteria or mononuclear cells from one compartment to the other. Mononuclear cells were placed on one side of the filter and *L. pneumophila* on the same side of the filter or on the opposite side of the filter from the mononuclear cells. The bacteria multiplied only when placed on the same side of the filter as mononuclear cells.[1]

Inside human monocytes, *L. pneumophila* multiplies with a doubling time at mid-log phase of just under 2 hr. This rate is at least as fast or faster than that reported for growth of *L. pneumophila* in artificial medium.

L. pneumophila is therefore an intracellular parasite. Since *L. pneumophila* can multiply extracellularly on complex media, this bacterium is a facultative intracellular parasite. *L. pneumophila* thus falls into a select category of bacterial pathogens that can evade host defenses by parasitizing mononuclear phagocytes. Other members of this category are the mycobacteria including *Mycobacterium tuberculosis* and *Mycobacterium leprae*, *Listeria monocytogenes*, *Brucella* sp., *Salmonella* sp., and *Francisella tularensis*.

III. A NOVEL INTRACELLULAR SITE OF REPLICATION

Inside human monocytes and alveolar macrophages, *L. pneumophila* multiplies in a novel cytoplasmic vacuole[1,2] (Figure 2). The cytoplasmic sides of the membrane-bound vacuole are studded with structures that are apparently host cell ribosomes. This morphologic feature is also found in infected alveolar macrophages in human lung tissue specimens obtained from patients with Legionnaires' disease.[9]

Formation of the *L. pneumophila* vacuole involves a complex series of cytoplasmic events that take place over 4 to 8 hr after entry of live *L. pneumophila* into human monocytes.[10]

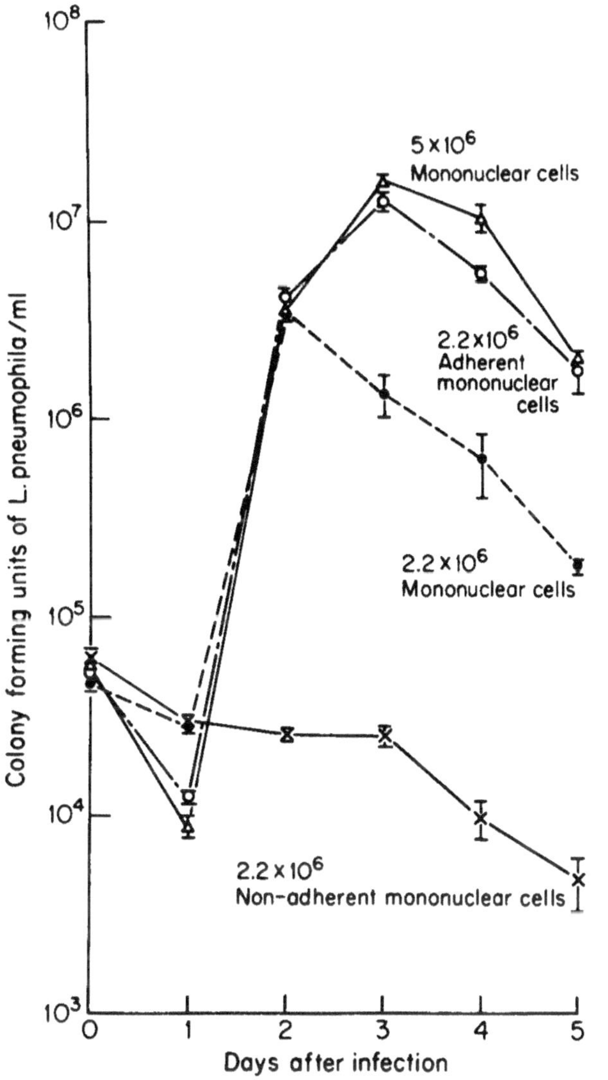

FIGURE 1. *L. pneumophila* multiplies in the presence of monocytes (adherent cells) but not lymphocytes (nonadherent cells). *L. pneumophila* (10^5 CFU) were added to petri dishes containing 2.2 mℓ RPMI, 15% serum, and (a) 2.2 or 5 × 10^6 mononuclear cells: (b) 2.2 × 10^6 adherent mononuclear cells selected from 5 × 10^6 mononuclear cells; (c) 2.2 × 10^6 nonadherent mononuclear cells selected from the same mononuclear cell population as in (b). The cultures were incubated at 37°C in 5% CO_2-95% air on a gyratory shaker (100 rpm) for 3 hr and under stationary conditions thereafter. CFU of *L. pneumophila* in each petri dish were determined daily. Each point represents the average for three replicate petri dishes ± SE. (Reproduced from Horwitz, M. A. and Silverstein, S. C., *J. Clin. Invest.*, 66, 441, 1980. With permission.)

Immediately after entry, no cytoplasmic organelles are present about the *L. pneumophila*-containing vacuole. By 15 min after entry, smooth vesicles surround the majority of such vacuoles. By 1 hr, mitochondria are found closely apposed to the membrane of a majority of the vacuoles. By 4 hr, the reaction of smooth vesicles and mitochondria about the vacuoles has abated, and ribosomes or ribosome-studded vesicles are found surrounding the vacuoles.

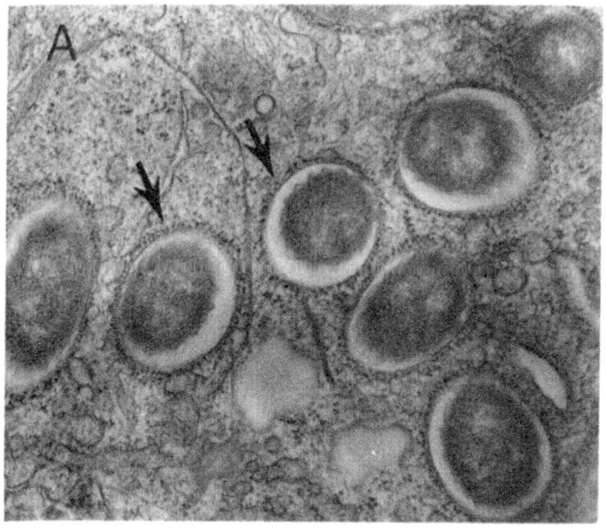

A.

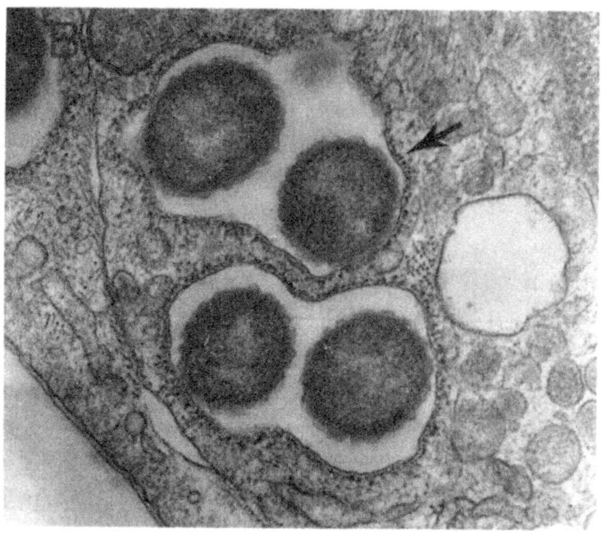

B.

FIGURE 2. Electron micrographs of human monocytes in monolayer culture infected with *L. pneumophila*. (A,B) *L. pneumophila* inside monocytes in membrane-bound cytoplasmic vacuoles studded with ribosomes (arrows). (A) Early after infection, each vacuole contains a single bacillus. (Magnification × 27,000.) (B) Each vacuole at a later stage of infection contains two bacilli. (Magnification × 32,400.) (C) Monocyte with a single large membrane-bound vacuole containing many *L. pneumophila*. Several bacteria appear to be in the process of dividing by binary fission. Although not apparent at this magnification, this vacuole is seen studded with ribosomes at higher magnification. (Magnification × 4,500.) (D) Monocyte at late stage of infection packed full with *L. pneumophila*. (Magnification × 5,400.) (Band D reproduced from Horwitz, M. A. and Silverstein, S. C., J. Clin. Invest., 66, 441, 1980. With permission.

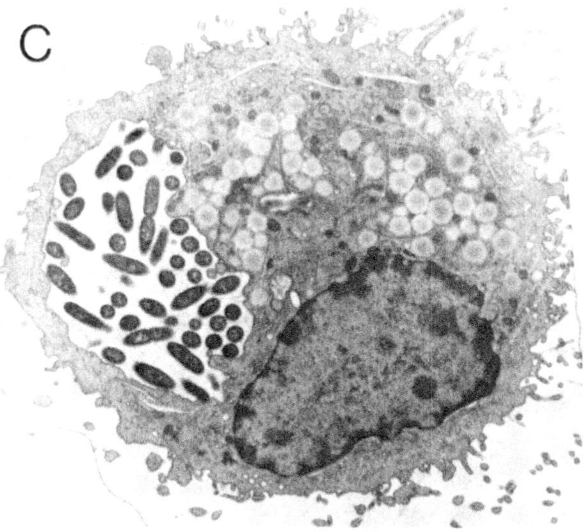

Figure 2C

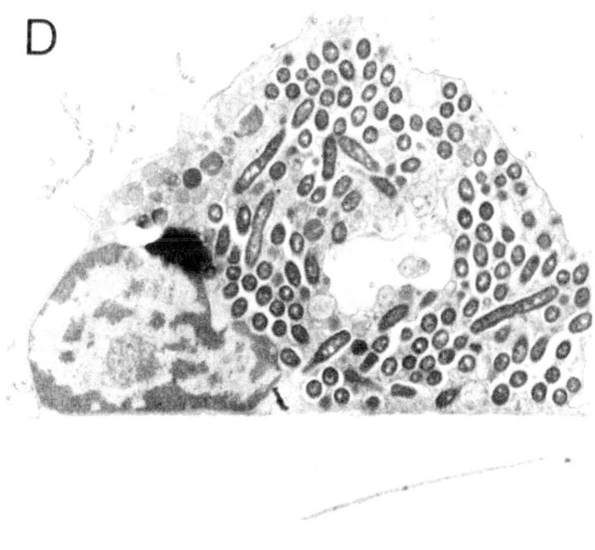

Figure 2D

By 8 hr, the vacuoles are studded with ribosomes.[10] Erythromycin at concentrations that inhibit the intracellular multiplication of *L. pneumophila* does not inhibit this sequence of events.[10] Formalin-killed *L. pneumophila* also enter monocytes and reside in membrane-bound cytoplasmic vacuoles; however, in contrast to the situation with live *L. pneumophila*, no reaction of cytoplasmic organelles comparable to that just described takes place about vacuoles containing formalin-killed *L. pneumophila* at any time after entry. Formalin-killed *L. pneumophila* are digested by monocytes within about 4 hr.[10]

IV. INTERACTION BETWEEN *L. PNEUMOPHILA* AND HUMAN COMPLEMENT AND ANTIBODY

Patients with Legionnaires' disease respond to the infection by producing antibody against *L. pneumophila*. Antibody might function to promote host defense in three major ways. First, it might promote killing of *L. pneumophila* by complement. Second, it might promote phagocytosis and killing of *L. pneumophila* by polymorphonuclear leukocytes (PMN) and mononuclear phagocytes. Third, it might inhibit the intracellular multiplication of *L. pneumophila* in mononuclear phagocytes. These potential functions of antibody will be considered in this and the following two sections of this chapter.

In one study, virulent egg yolk-grown *L. pneumophila* Philadelphia 1 were found to be completely resistant to the bactericidal effects of human complement.[8] In another study, *L. pneumophila* Knoxville 1 and Bellingham 1 grown on Mueller-Hinton agar supplemented with hemoglobin and cysteine and therefore presumably avirulent[11] were found to be sensitive to human complement.[12] The disparate results of these studies are likely related to the different methods of culturing the bacteria. If so, then *L. pneumophila* that are pathogenic for man are likely serum resistant, since, in their virulence, they more closely resemble egg yolk-grown than agar-adapted bacteria.

In the presence of high-titer human or rabbit anti-*L. pneumophila* antibody, virulent *L. pneumophila* Philadelphia 1 remain completely resistant to the bactericidal effects of human serum.[8]

The resistance of *L. pneumophila* to complement is not due to a failure to fix complement.[8] Fluorescence microscopy using rhodamine-conjugated anti-human complement C3 IgG has demonstrated that complement component C3 is fixed to the bacterial surface in the presence of antibody; however, fixation of complement does not result in death of the bacterium.[8]

V. INTERACTION BETWEEN *L. PNEUMOPHILA* AND HUMAN POLYMORPHONUCLEAR LEUKOCYTES

Human polymorphonuclear leukocytes (PMN) require both anti-*L. pneumophila* antibody and complement to ingest live virulent *L. pneumophila* efficiently.[8] In the absence of either antibody or complement, PMN bind or ingest 95% fewer *L. pneumophila* than in the presence of both serum factors.[8] Inside PMN, opsonized *L. pneumophila* are found in membrane-bound cytoplasmic vacuoles.[8]

Paralleling their requirements for phagocytosis of *L. pneumophila*, human PMN require both anti-*L. pneumophila* antibody and complement to kill any *L. pneumophila*.[8] Even then, PMN kill only 0.5 log of an inoculum.[8]

The experiment depicted in Figure 3 illustrates the remarkable difference in the capacity of PMN to kill *L. pneumophila* and an encapsulated serum-resistant strain of *Escherichia coli*. In this experiment, PMN were incubated with *L. pneumophila* alone, *E. coli* alone, or both types of bacteria at the same time. PMN killed few *L. pneumophila* even in the presence of high-titer anti-*L. pneumophila* antibody and fresh serum. In contrast, under these same conditions, PMN very effectively killed an encapsulated serum-resistant strain of *E. coli*, reducing the number of viable bacteria by 2-1/2 logs. (PMN require both antibody and complement to kill encapsulated *E. coli*;[13] both are present in fresh normal human serum, which contains low levels of natural antibody against this strain of *E. coli*.) When the *E. coli* and *L. pneumophila* were mixed together in the same test tube, PMN killed the *E. coli* as effectively as when the *E. coli* were present alone, indicating that the killing capacity of PMN is not reduced in cultures containing *L. pneumophila*.

L. pneumophila surviving incubation with PMN in the presence of antibody and complement are PMN-associated.[8] Therefore, the relatively ineffective killing of opsonized *L.*

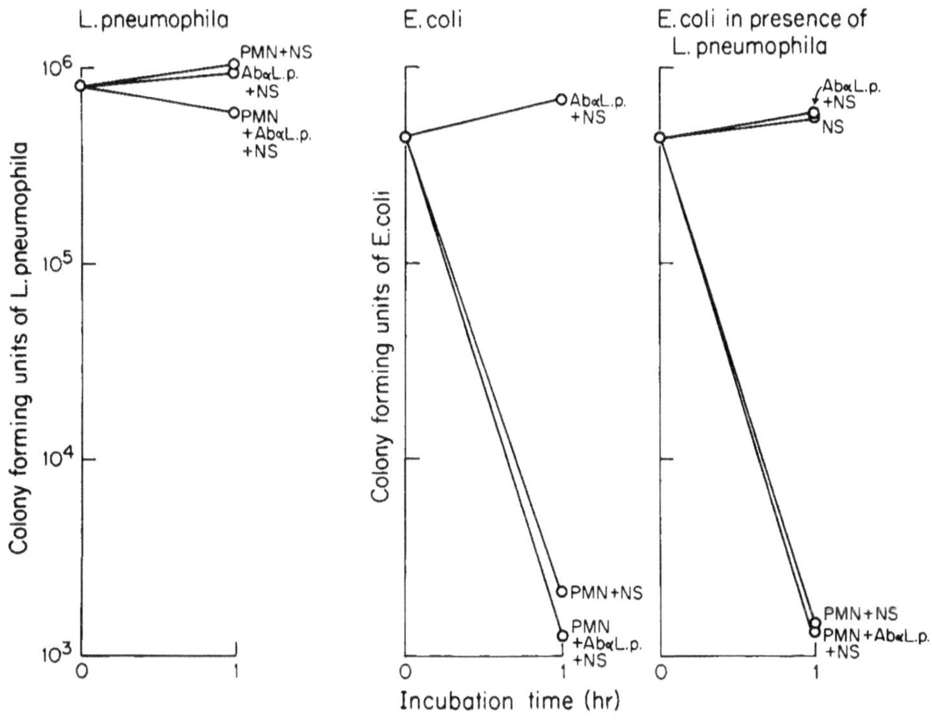

FIGURE 3. *L. pneumophila* resists killing by PMN under conditions in which PMN effectively kill a serum-resistant encapsulated strain of *E. coli*. PMN (2.5×10^6) and 5×10^5 CFU of *L. pneumophila* and/or *E. coli* were incubated at 37°C for 1 hr or a gyratory shaker in 0.9 mℓ medium (final volume) containing 10% fresh normal human serum (NS) alone or 10% NS plus 8.5 agglutinating units/mℓ of dialyzed rabbit anti-*L. pneumophila* antiserum (Abα-L.p.). CFU of *E. coli* and *L. pneumophila* were determined initially and at the end of the incubation. (Reproduced from Horwitz, M. A. and Silverstein, S. C., *J. Exp. Med.*, 153, 386, 1981. With permission.)

pneumophila by PMN is a result of a failure of PMN to kill these bacteria after they become PMN-associated.[6]

Thus, in the presence of both antibody and complement, PMN efficiently phagocytize *L. pneumophila*, but they kill a relatively modest proportion of an inoculum.[8]

VI. INFLUENCE OF ANTIBODY ON THE INTERACTION BETWEEN *L. PNEUMOPHILA* AND MONONUCLEAR PHAGOCYTES

Human monocytes, like human PMN, phagocytize virulent *L. pneumophila* much more efficiently in the presence of both human anti-*L. pneumophila* antibody and human complement than in the presence of human complement alone.[14] Similarly, mouse peritoneal macrophages and rat alveolar macrophages are reported to phagocytize broth-grown *L. pneumophila* more efficiently in the presence of rabbit antiserum raised against the homologous serotype than in the presence of rabbit antiserum raised against a heterologous serotype.[15] Also guinea pig macrophages are reported to phagocytize virulent *L. pneumophila* more efficiently in the presence of immune guinea pig serum then in the presence of nonimmune guinea pig serum.[5]

Human monocytes, again like human PMN, require both anti-*L. pneumophila* antibody and complement to kill any *L. pneumophila*, but even then, the monocytes kill only a limited proportion (0.25 to 0.5 log) of an inoculum.[14] *L. pneumophila* surviving the initial encounter with monocytes multiply in the monocytes and at as rapid a rate as bacteria that enter monocytes in the absence of antibody[14] (Figure 4).

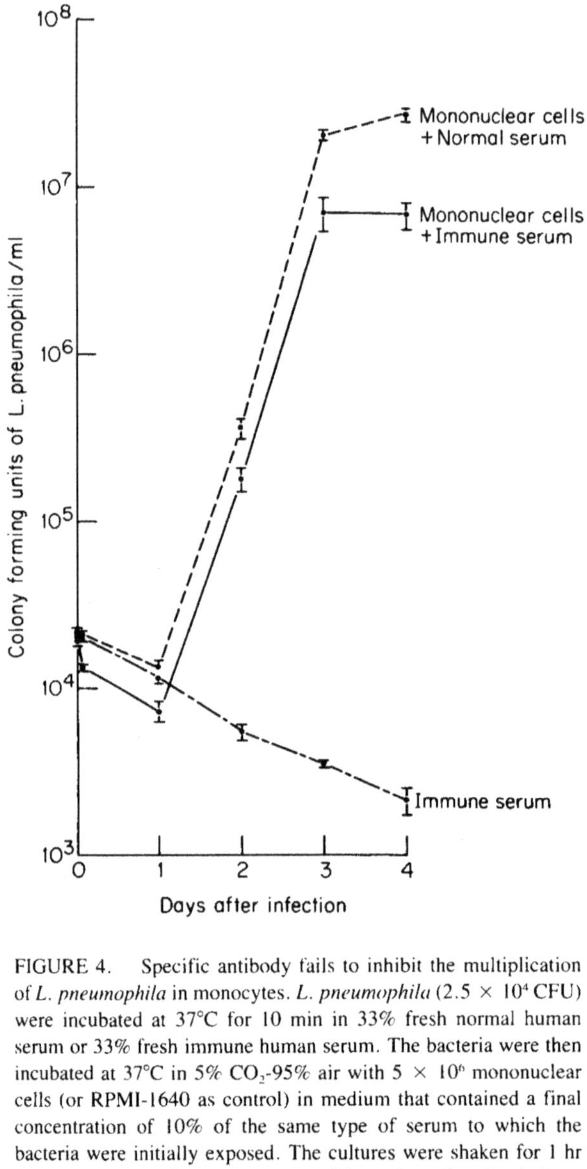

FIGURE 4. Specific antibody fails to inhibit the multiplication of *L. pneumophila* in monocytes. *L. pneumophila* (2.5×10^4 CFU) were incubated at 37°C for 10 min in 33% fresh normal human serum or 33% fresh immune human serum. The bacteria were then incubated at 37°C in 5% CO_2-95% air with 5×10^6 mononuclear cells (or RPMI-1640 as control) in medium that contained a final concentration of 10% of the same type of serum to which the bacteria were initially exposed. The cultures were shaken for 1 hr and incubated under stationary conditions thereafter for 4 days. CFU were determined at 0, 1, 24, 48, 72, and 96 hr after infecting the monocytes. Each point represents the average for five replicate tubes ±SE. (Reproduced from Horwitz, M. A. and Silverstein, S. C., *J. Exp. Med.*, 153, 398, 1981. With permission.

Thus three major ways in which antibody and complement might function to promote host defense have been investigated. These studies reveal (a) that antibody fails to promote killing of virulent *L. pneumophila* by complement; (b) that antibody and complement together fail to promote effective killing of *L. pneumophila* by PMN and monocytes; and (c) that antibody and complement fail to inhibit the rate of intracellular multiplication of *L. pneumophila* in monocytes. These findings suggest that humoral immunity may not be an effective host defense against *L. pneumophila* and that a vaccine that resulted only in antibody production against *L. pneumophila* may not be efficacious.[8,14]

VII. INTERACTION BETWEEN *L. PNEUMOPHILA* AND ACTIVATED HUMAN MONONUCLEAR PHAGOCYTES

Cell-mediated immunity plays an important role in host defense against intracellular pathogens. Lymphokines produced by sensitized lymphocytes have been reported to activate animal or human mononuclear phagocytes such that they have an enhanced capacity to resist or kill *M. tuberculosis, L. monocytogenes, Leishmania enrietti, Toxoplasma gondii, Trypanosoma cruzi, Rickettsia tsutsugamushi,* and *Coxiella burnetti.*[16] The interaction between in vitro activated human mononuclear phagocytes and *L. pneumophila* has also been examined.[2,17]

Freshly explanted normal human monocytes activated by incubation with concanavalin A and human lymphocytes inhibit the intracellular multiplication of *L. pneumophila.*[17] Both Con A and lymphocytes are required for activation under these conditions[17] (Figure 5).

Normal monocytes activated by incubation with cell-free filtered supernatants from concanavalin A stimulated mononuclear cell cultures also inhibit the intracellular multiplication of *L. pneumophila* and the degree of inhibition is proportional to the concentration of supernatant added[17] (Figure 6). The most potent supernatants are obtained from mononuclear cell cultures incubated with ≥15 μg/mℓ Con A for 48 hr.[17] The degree of monocyte inhibition of *L. pneumophila* multiplication is proportional to the length of time monocytes are preincubated with supernatant (48 > 24 > 12 hr) and monocytes treated with supernatant daily after infection are more inhibitory than monocytes treated initially only.[17] With time in culture, monocytes progressively lose a limited degree of spontaneous inhibitory capacity and their capacity to respond to supernatant with inhibition of *L. pneumophila* multiplication.[17]

Human alveolar macrophages, activated in vitro by incubation with cell-free supernatants of concanavalin A-stimulated mononuclear cell cultures, also inhibit the intracellular multiplication of *L. pneumophila.*[2]

Supernatant-activated monocytes inhibit *L. pneumophila* multiplication in two ways. First, they phagocytize approximately 50% fewer bacteria than nonactivated monocytes and thereby restrict access of these bacteria to the intracellular milieu they require for multiplication.[17] Second, activated monocytes markedly slow the multiplication rate of bacteria that are internalized; the doubling time of *L. pneumophila* in activated monocytes during the first 22 hr after infection was 3.1 times longer than in nonactivated monocytes.[17]

Activated monocytes do not have a greater capacity to kill *L. pneumophila* than nonactivated monocytes.[17] In the presence of fresh normal serum, containing complement but not antibody, monocytes do not kill *L. pneumophila* whether or not the monocytes are activated.[17] In the presence of fresh immune serum, containing both anti-*L. pneumophila* antibody and complement, both activated and nonactivated monocytes kill a small proportion (<0.25 log) of an inoculum of *L. pneumophila.*[17] The activated monocytes do not kill more *L. pneumophila*; in fact, they consistently kill fewer bacteria than the nonactivated monocytes, perhaps because they phagocytize less avidly.[17]

These experiments show that human monocytes and alveolar macrophages can be activated to inhibit the multiplication of *L. pneumophila.* Thus, inhibition of *L. pneumophila* multiplication is accomplished by activating mononuclear phagocytes and not by coating the bacteria with antibody and complement. This indicates that cell-mediated immunity may play an important role in host defense against *L. pneumophila.*

VIII. EVIDENCE FOR CELL-MEDIATED IMMUNITY TO *L. PNEUMOPHILA* IN HUMANS

Evidence that humans develop cell-mediated immunity to *L. pneumophila* has been obtained from in vitro studies.[18] Mononuclear cells from patients recovered from Legionnaires'

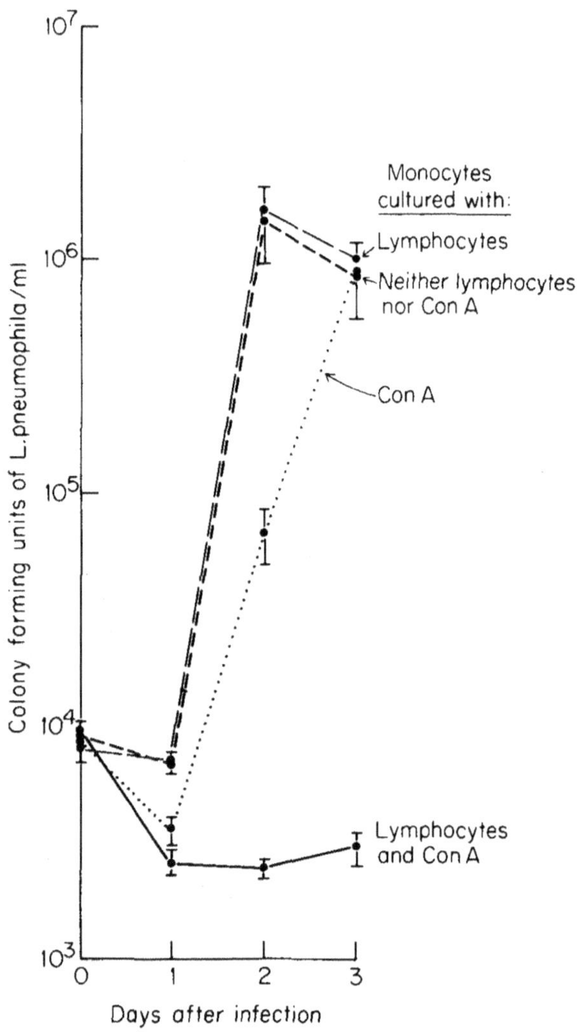

FIGURE 5. Monocytes incubated with both Con A and lymphocytes inhibit *L. pneumophila* multiplication. Mononuclear cells (6×10^6) in 2 mℓ medium containing 25% fresh human serum were incubated in 35 mm plastic petri dishes for 1.5 hr at 37°C in 5% CO_2-95% air to allow monocytes to adhere. Some cultures were vigorously washed to remove nonadherent (predominantly lymphoid) cells. These and the other cultures were then further incubated for 24 hr with or without Con A (16 μg/mℓ). *L. pneumophila* (2×10^4 CFU) were then added to each culture and CFU in each culture determined daily. Each point represents the average for 3 replicate petri dishes ± SE. (Reproduced from Horwitz, M. A. and Silverstein, S. C., *J. Exp. Med.*, 154, 1618, 1981. With permission.)

disease respond to *L. pneumophila* antigens with proliferation (lympho-proliferation), as measured by the capacity of the mononuclear cells to incorporate ^{3}H-thymidine. Moreover, these mononuclear cells respond with the production of supernatants containing monocyte-activating cytokines; such cytokines have the capacity to activate normal monocytes such that they inhibit the intracellular multiplication of *L. pneumophila*.[18] The degree of inhibition is proportional to the concentration of supernatant added.[18] Maximally potent supernatants are produced by incubation of patient mononuclear cells with *L. pneumophila* antigens for

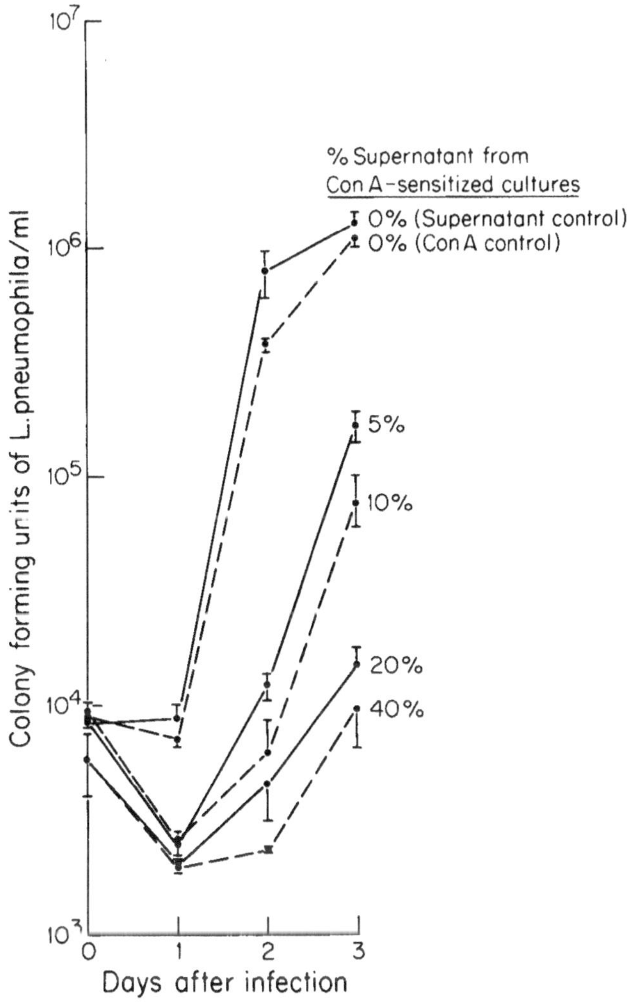

FIGURE 6. Monocytes incubated with the supernatant of Con A-sensitized mononuclear cell cultures inhibit *L. pneumophila* multiplication. Monocytes in monolayer culture were incubated at 37°C in 5% CO_2-95% air in 2 mℓ RPMI medium containing 15% fresh normal human serum and 0 to 40 % cell-free supernatant from Con A-sensitized mononuclear cell cultures. Control monolayers were incubated with supernatant from a mononuclear cell culture from which Con A was omitted (Supernatant control) or with 15 μg/mℓ Con A (Con A control). After 24 hr *L. pneumophila* (2×10^4 CFU) were added to the cultures and CFU in each culture determined daily. Each point represents the average for 3 replicate petri dishes ± SE. (Reproduced from Horwitz, M. A. and Silverstein, S. C., *J. Exp. Med.*, 154, 1618, 1981. With permission.)

4 days.[18] Mononuclear cells from persons without historical or serological evidence of *L. pneumophila* infection do not respond strongly to *L. pneumophila* antigens in these assays.[18]

Thus cell-mediated immunity to *L. pneumophila* develops in patients with Legionnaires' disease. The studies on the interaction between activated human mononuclear phagocytes and *L. pneumophila* described above indicate that cell-mediated immunity plays a major role in host defense against *L. pneumophila* as it does against other intracellular pathogens.

In vivo studies in guinea pigs also indicate a role for cell-mediated immunity in *L. pneumophila* infection. Guinea pigs sensitized by injection of *L. pneumophila* antigens exhibit a cutaneous delayed-type hypersensitivity reaction in response to *L. pneumophila* antigens.[19]

IX. INFLUENCE OF ERYTHROMYCIN AND RIFAMPIN ON *L. PNEUMOPHILA* LOCATED INTRACELLULARLY IN HUMAN MONOCYTES

Erythromycin and rifampin are currently the drugs of choice for the treatment of Legionnaires' disease. Both clinical experience and in vivo studies indicate that these antibiotics are efficacious in the treatment of Legionnaires' disease.[20-23] The influence of antibiotics on *L. pneumophila* multiplying extracellularly under cell-free conditions on artificial medium has been well studied;[24,25] erythromycin and rifampin inhibit the extracellular multiplication of *L. pneumophila* and kill bacteria multiplying extracellularly at relatively low concentrations of antibiotic.[24-26] Since *L. pneumophila* multiply intracellularly in mononuclear phagocytes, it is important to understand the influence of antibiotics on intracellular *L. pneumophila*.

The intracellular multiplication of *L. pneumophila* in human monocytes is inhibited by concentrations of erythromycin and rifampin comparable to those which inhibit extracellular multiplication of *L. pneumophila*.[26] The antibiotics inhibit intracellular multiplication of *L. pneumophila* whether added just before or just after infection of monocytes, or 2 days after infection when the bacteria are in the logarithmic phase of growth in monocytes.[26] Multiplication of *L. pneumophila* in the logarithmic phase of growth in monocytes is inhibited within 1 hr of the addition of antibiotics.[26] Intracellular bacteria inhibited from multiplying by antibiotics are not killed. By electron microscopy, the bacteria appear intact within membrane-bound vacuoles, studded with ribosome-like structures.[26]

As mentioned, *L. pneumophila* multiplying extracellularly on artificial medium are killed readily by relatively low concentrations of erythromycin and rifampin; the MBC is 1 μg/mℓ for erythromycin and 0.009 μg/mℓ for rifampin.[26] In contrast, *L. pneumophila* multiplying intracellularly are resistant to killing by these concentrations of erythromycin and rifampin or by concentrations equal to or greater than peak serum levels in humans, levels 12 times the MBC for erythromycin and 10,000 times the MBC for rifampin.[26] Extracellular *L. pneumophila* in stationary phase are also resistant to killing by erythromycin and rifampin.[26] This suggests that, in vivo, *L. pneumophila* are resistant to killing by erythromycin and rifampin.

Inhibition of *L. pneumophila* multiplication in monocytes by antibiotics is reversible; when the antibiotics are removed from infected monocyte cultures after 2 days, *L. pneumophila* resume multiplying.

These findings indicate that patients with Legionnaires' disease under treatment with erythromycin and rifampin require host defenses to eliminate *L. pneumophila* and that inadequate host defenses may result in relapse after cessation of therapy.

X. INTRACELLULAR SURVIVAL OF *L. PNEUMOPHILA*: INHIBITION OF PHAGOSOME-LYSOSOME FUSION

Some intracellular pathogens, such as *Toxoplasma gondii, Mycobacterium tuberculosis, Mycobacterium microti,* and *Chlamydia psittaci* inhibit fusion of phagosomes with lysosomes.[16] Others, such as *Leishmania* and *Mycobacterium lepraemurium* do not inhibit fusion, and they apparently are able to survive within the normally inhospitable milieu of the phagolysosome.[16] *L. pneumophila* falls in the first category of organisms.[10] Live *L. pneumophila* inhibit fusion of the phagosomes with both primary and secondary lysosomes of monocytes.[10] In contrast, formalin-killed *L. pneumophila* do not inhibit fusion of the phagosome with monocyte primary or secondary lysosomes.[10] These findings suggest that inhibition of fusion may be an important mechanism for survival of *L. pneumophila* in phagocytes.

Further evidence that inhibition of phagosome-lysosome fusion may play a role in *L. pneumophila* survival comes from studies of the susceptibility of *L. pneumophila* to oxygen metabolites generated by phagocytes. In a leukocyte-free system, *L. pneumophila* is susceptible to moderate concentrations of hydrogen peroxide (LD_{50} = 1 to 5 × 10^{-4}(M).[27] However, the susceptibility of *L. pneumophila* to hydrogen peroxide is increased 100-fold by the addition of lactoperoxidase and a halide.[27] (Peroxidase and a halide have been shown to increase the susceptibility of numerous microorganisms to hydrogen peroxide.[28]) This suggests that one way by which inhibition of phagosome-lysosome fusion may enhance the survival of *L. pneumophila* in monocytes or polymorphonuclear leukocytes is by blocking access of lysosomal or granule myeloperoxidase to the phagosome. Inhibition of phagosome-lysosome fusion may also block access of other lysosomal anti-microbial substances to the phagosome.

L. pneumophila has been reported susceptible to other anti-microbial oxygen metabolites that leukocytes may generate. The bacterium is killed by products of the acetaldehyde-xanthine oxidase system which generates superoxide, hydrogen peroxide, and hydroxyl radicals.[29] *L. pneumophila* appears primarily susceptible to hydroxyl radical in this system.[29]

XI. CONCLUSIONS

L. pneumophila, an organism that is ubiquitous in aquatic environments, has acquired the capacity to survive and multiply within human mononuclear phagocytes. The ubiquitousness of *L. pneumophila* indicates that man is an incidental host and that the organism multiplies in nature in other hosts, perhaps amoebae, or in other ways, perhaps extracellularly. It is interesting to speculate further that amoebae may have provided the evolutionary training ground where *L. pneumophila* acquired its capacity to thrive intracellularly and that a common mechanism underlies the capacity of this bacterium to survive in cells so distantly related, evolutionarily speaking, as amoebae and mononuclear phagocytes. In any case the mechanism by which *L. pneumophila* survives intracellularly in mononuclear phagocytes remains to be determined, although the bacterium's capacity to inhibit phagosome lysosome fusion would appear to play a central role.

Inside mononuclear phagocytes, *L. pneumophila* multiplies in a novel membrane-bound cytoplasmic vacuole studded with ribosomes. How *L. pneumophila* orchestrates the complex sequence of events culminating in formation of this vacuole and the role of this vacuole in the bacterium's intracellular life remain unknown. The search for answers to these puzzles may yield important information about intracellular parasitism and the physiology of mononuclear phagocytes.

Cell-mediated immunity appears to play a primary role in host defense against *L. pneumophila*. Patients with Legionnaires' disease develop cell-mediated immunity to *L. pneumophila*; their mononuclear cells respond to *L. pneumophila* antigens with the production of cytokines capable of activating monocytes such that they inhibit the multiplication of *L. pneumophila*. Humoral immunity appears to play a modest role in host defense against *L. pneumophila*. Phagocytes, in conjunction with antibody and complement, phagocytize and kill a limited proportion of an inoculum of *L. pneumophila*. Thus, humoral immunity may assist the host in eliminating *L. pneumophila* from the body once cell-mediated immunity has been established and the intracellular multiplication of *L. pneumophila* has been inhibited.

ACKNOWLEDGMENTS

This work was supported by the American Cancer Society, the John A. Hartford Foundation, National Institutes of Health grants AI 17254, and CA 30198 and the Heiser Foundation for Research in Leprosy.

REFERENCES

1. **Horwitz, M. A. and Silverstein, S. C.,** The Legionnaires' disease bacterium (*Legionella pneumophila*) multiplies intracellularly in human monocytes, *J. Clin. Invest.*, 66, 441, 1980.
2. **Nash. T. W., Libby, D. M., and Horwitz, M. A.,** Legionnaires' disease bacteria multiply in human alveolar macrophages and activated alveolar macrophages inhibit multiplication, in Program of the Twenty-second Interscience Conference on Antimicrobial Agents and Chemotherapy, sponsored by the American Society for Microbiology, Miami Beach, Fla., October 4 to 6, 1982.
3. **Kishimoto, R. A., Kastello, M. D., White, J. D., Shirey, F. G., McGann, V. G., Larson, E. W., and Hedlund, K. W.,** In vitro interaction between normal cynolmolgus monkey alveolar macrophages and Legionnaires' disease bacteria, *Infect. Immun.*, 25, 761, 1979.
4. **Jacobs, R. F., Locksley, R. M., Wilson, C. B., Haas, J. E., and Klebanoff, S. J.,** Effects of alveolar macrophages on *Legionella pneumophila*, *Clin. Res.*, 30, 431A, 1982.
5. **Kishimoto, R. A., White, J. D., Shirey, F. G., McGann, V. G., Berendt, R. F., Larson, E. W., and Hedlund, K. W.,** In vitro response of guinea pig peritoneal macrophages to *Legionella pneumophila*, *Infect. Immun.*, 31, 1209, 1981.
6. **Daisy, J. A., Benson, C. E., McKitrick, J., and Friedman, H. M.,** Intracellular replication of *Legionella pneumophila* in cultured human embryonic lung fibroblasts, *Infect. Immun.*, 28, 1014, 1980.
7. **Wong, M. C., Ewing, E. P., Callaway, C. S., and Peacock, W. L., Jr.,** Intracellular multiplication of *Legionella pneumophila* in cultured human embryonic lung fibroblasts, *Infect. Immun.*, 28, 1014, 1980.
8. **Horwitz, M. A. and Silverstein, S. C.,** Interaction of the Legionnaires' disease bacterium (*Legionella pneumophila*) with human phagocytes. I. *L. pneumophila* resists killing by polymorphonuclear leukocytes, antibody, and complement, *J. Exp. Med.*, 153, 386, 1981.
9. **Glavin, F. L., Winn, W. C., Jr., and Craighead, J. E.,** Ultrastructure of lung in Legionnaires' disease. Observations of three biopsies done during the Vermont epidemic, *Ann. Intern. Med.*, 90, 555, 1979.
10. **Horwitz, M. A.,** Interaction between Legionnaires' disease bacteria and human monocytes: formation of a unique intracellular vacuole and inhibition of phagosome — lysosome fusion, in Program of the Twenty-second Interscience Conference on Antimicrobial Agents and Chemotherapy, sponsored by the American Society for Microbiology, Miami Beach, Fla., October 4 to 6, 1982.
11. **McDade, J. E. and Shepard, C. C.,** Virulent to avirulent conversion of Legionnaires' disease bacterium (*Legionella pneumophila*) — its effect on isolation techniques, *J. Infect. Dis.*, 139, 707, 1979.
12. **Arko, R. J., Wong, K. H., and Feeley, J. C.,** Immunologic factors affecting the in vivo and in vitro survival of the Legionnaires' disease bacterium, *Ann. Int. Med.*, 90, 680, 1979.
13. **Horwitz, M. A. and Silverstein, S. C.,** Influence of the *Escherichia coli* capsule on complement fixation and on phagocytosis and killing by human phagocytes, *J. Clin. Invest.*, 65, 82, 1980.
14. **Horwitz, M. A. and Silverstein, S. C.,** Interaction of the Legionnaires' disease bacterium (*Legionella pneumophila*) with human phagocytes. II. Antibody promotes binding of *L. pneumophila* to monocytes but does not inhibit intracellular multiplication, *J. Exp. Med.*, 153, 398, 1981.
15. **Johnson, W., Pesanti, E., and Elliot, J.,** Serospecificity and opsonic activity of antisera to *Legionella pneumophila*, *Infect. Immun.*, 26, 698, 1979.
16. **Horwitz, M. A.,** Phagocytosis of microorganisms, *Rev. Infect. Dis.*, 4, 104, 1982.
17. **Horwitz, M. A. and Silverstein, S. C.,** Activated human monocytes inhibit the intracellular multiplication of Legionnaires' disease bacteria, *J. Exp. Med.*, 154, 1618, 1981.
18. **Horwitz, M. A.,** Demonstration of cell-mediated immunity in Legionnaires' disease, *Clin. Res.*, 30, 369A, 1982.
19. **Wong, K. H., McMaster, P. R. B., Feeley, J. C., Arko, R. J., Schalla, W. O., and Chandler, F. W.,** Detection of hypersensitivity to *Legionella pneumophila* in guinea pigs by skin test, *Curr. Microbiol.*, 4, 105, 1980.
20. **Fraser, D. W., Tsai, T. R., Orenstein, W., Parkin, W. E., Beecham, H. J., Sharrar, P. G., Harris, J., Mallison, G. F., Martin, S. M., McDade, J. E., Shepard, C. C., Brachman, P. S., and the Field Investigation Team,** Legionnaires' disease. Description of an epidemic of pneumonia, *N. Engl. J. Med.*, 297, 1189, 1977.
21. **Meyer, R. P. and Finegold, S. M.,** Legionnaires' disease, *Ann. Rev. Med.*, 31, 219, 1980.
22. **Fraser, D. W., Bopp, C., Wachsmuth, I. K., Feeley, J. C., and Tsai, T.,** Antibiotic treatment of guinea-pigs infected with agent of Legionnaires' disease, *Lancet*, 1, 175, 1978.
23. **Lewis, V. J., Thacker, W. L., Shepard, C. C., and McDade, J. E.,** In vivo susceptibility of the Legionnaires' disease bacterium to ten antimicrobial agents, *Antimicrob. Agents Chemother.*, 13, 419, 1978.
24. **Thornsberry, C., Baker, C. N., and Kirven, L. A.,** In vitro activity of antimicrobial agents on Legionnaires' disease bacterium, *Antimicrob. Agents Chemother.*, 13, 78, 1978.
25. **Thornsberry, C. and Kirven, L. A.,** β-Lactamase of the Legionnaires' bacterium, *Curr. Microbiol.*, 1, 51, 1978.

26. **Horwitz, M. A. and Silverstein, S. C.,** The intracellular multiplication of Legionnaires' disease bacteria (*Legionella pneumophila*) in human monocytes is reversibly inhibited by erythromycin and rifampin, *J. Clin Invest.*, 71, 15, 1983.
27. **Horwitz, M. A.,** Unpublished data.
28. **Klebanoff, S. J.,** Oxygen metabolism and the toxic properties of phagocytes, *Ann. Int. Med.*, 93, 480, 1980.
29. **Locksley, R. M., Jacobs, R. F., Wilson, C. B., and Klebanoff, S. J.,** Susceptibility of *Legionella pneumophila* to oxygen-dependent microbicidal systems, *Clin. Res.*, 30, 519A, 1982.

Chapter 10

IMMUNITY TO LEGIONELLA

Toby K. Eisenstein and Herman Friedman

TABLE OF CONTENTS

I. INTRODUCTION

A. Host-Pathogen Interactions

The definition of the offensive and defensive strategies of the host and the pathogen are a subject of unending fascination for the student of infectious diseases. Whenever a new combatant appears to challenge our immune system, we are intellectually driven to seek the means by which it has penetrated our host defense mechanisms. The knowledge gleaned from our unveiling of the microscopic battlefield has both basic and applied objectives; the basic being the satisfaction of exposing the game-plan of the intruder, and the applied, the promise of development of effective vaccines and diagnostic tests.

With the discovery that Legionnaires' disease was caused by a previously unrecognized bacterium, we could begin to dissect the cellular and molecular events peculiar to the struggle of this organism with its host, and to compare its mechanisms of pathogenesis and immunity with strategies already documented for other bacteria. There are three major mechanisms by which bacteria are thought to cause disease. These are (1) intoxication, (2) evasion of phagocytosis with unchecked extracellular multiplication (septicemia or suppurative infection), and (3) intracellular replication, particularly within phagocytes, presumably by subversion of bactericidal mechanisms of the phagocyte. Immunity to organisms using strategies one and two is mainly humoral, i.e., antibody to the toxin is neutralizing, or antibody to the organism is opsonic and allows the polymorphonuclear leukocyte to ingest, and thus dispose of it. Immunity to intracellular pathogens, however, seems to require participation of lymphocytes of the thymus-dependent arm of the immune system, which when exposed to microbial antigens to which they have already been sensitized, elaborate lymphokines that in turn activate macrophages to enhanced bactericidal or bacteriostatic capacity.[1,2] This T cell-macrophage effector pathway has been termed "cellular immunity." Prototypic organisms for each of the aforementioned strategies of subversion of the host are *Clostridium tetani* (intoxication), *Streptococcus pneumoniae* (extracellular replication), and *Mycobacterium tuberculosis* (intracellular pathogen). Certainly there are organisms which have in their armamentarium more than one of these strategies; for example, *Pseudomonas aeruginosa* is both toxigenic and extracellular, while antibody is protective against the facultative intracellular pathogen, *Toxoplasma gondii*.[3] In this regard, *Legionella* of serogroup 1 have been reported to have an extracellular cytotoxin[4,5] and also to exhibit endotoxic activity.[5,6] Clinical reports of neurological abnormalities in infected patients are consistent with the interpretation of a toxic syndrome in vivo.[6]

It is important to be able to categorize a new organism with regard to its mechanisms of pathogenesis, since vaccine strategies for induction of antibody-mediated immunity are quite different from those designed to result in cellular immunity. Antibodies can generally be induced by living or killed organisms, or purified bacterial extracts or toxoids. However, cellular immunity is thought to be induced only by living organisms or by microbial extracts suspended in Freund's Complete Adjuvant (FCA). Thus, for some intracellular pathogens, only attenuated vaccines are believed to be effective.

If, however, nonliving preparations can be shown to be protective for a certain organism, then the problem becomes one of defining the nature of the protective antigen(s) and of preparing it in a nontoxic form for use in a vaccine. In the case of *Legionella*, it is already clear that there are multiple serogroups as well as different species. Thus, if nonviable immunogens can be used, it will be necessary to either have a polyvalent vaccine, or to elucidate a common protective antigen. If attenuated vaccines are required, it will remain to be established whether cross-serogroup or cross-species protection can be induced by immunization with a single strain.

An unambiguous definition of which pathogenic strategy a microbe is using, and thus of which vaccination strategy is most likely to be successful, is not always easy. From clinical studies we can learn something about the major surface antigens of the organism(s) by examining the humoral response in convalescent sera. However, the mere presence of antibodies to an organism, while potentially useful in serotyping or in clinical diagnosis, gives no indication of whether the antibodies are protective. Thus, in tuberculosis infection a variety of antibodies to the organism, of differing specificities, can be detected, but none have been proven to have any role in host defense.[7] If a suitable skin test antigen for delayed hypersensitivity can be prepared, then convalescent and normal populations can be screened for reactivity, and such reactivity correlated with clinical or serological evidence of infection or prior exposure to the organism. Intracellular-type organism more frequently induce delayed-type hypersensitivity than toxigenic or extracellular pathogens.[8]

In clinical situations it has been observed that *Legionella* infect immuno-compromised hosts, although no particular immunological lesion has been associated with acquisition of the disease.[9] A recent report of infection with *Legionella pneumophila*, serogroup 6, in a 5-month-old baby with a severe combined immunodeficiency illustrates that individual cases are frequently found in patients with underlying disease and immunosuppression.[10] Perhaps closer scrutiny of compromised patients might reveal a correlation between selective T cell or B cell dysfunction and disease. However, to date no such relationship has emerged.

Evidence to delineate whether protective immunity is humoral or cellular is frequently aided by development of a suitable animal model or in vitro test system. A crucial aspect of evaluating such studies, however, is the accumulation of convincing evidence that the model chosen is valid, in that it accurately reflects the mode of pathogenesis in man. This point is far from trivial, for while it is possible to infect animals other than the natural host, the host-pathogen interactions may be quite different in the experimental situation. Thus, results from model systems must be interpreted with caution.

B. Animal Models for Studying *Legionella*

Legionella pneumophila was originally isolated by intraperitoneal infection of the guinea pig with biopsy and autopsy material from patients.[11] The sensitivity of the guinea pig to infection certainly makes it a prime candidate for a successful, and potentially valid animal model. Although the natural route of infection of man is thought to be by aerosol, the lung pathology of guinea pigs inoculated intraperitoneally does not differ significantly from that observed in pigs receiving organisms intranasally, and in both routes of infection, the histological findings are similar to those seen in patients.[12] Similarly, the pathological process in the lungs of guinea pigs infected by those routes of inoculation does not differ appreciably from that observed after intratracheal instillation of organisms[13] or by aerosol.[14] The intraperitoneal, aerosol, and intratracheal routes of infection have all been shown to produce fatal disease accompanied by at least some of the following signs of a true infection: weight loss, fever, pneumonia, bacterial multiplication in vivo, and antibody responses.[13,14,15] Thus, the guinea pig seems to be an excellent model for studying *Legionella*, even by routes other than aerosol.

Mice and rats are relatively resistant to *Legionella* by the intraperitoneal route of infection. Studies in the rat using intraperitoneal infection with mucin-coated organisms showed death within 24 hr of a large bacterial inoculum (6 to 9 $\times$ 10^8 *Legionella*). Although the lung pathology was not consistent with the human disease, the investigators interpreted their model as representative of a toxic syndrome.[16] AKR/J mice infected i.p. also succumb within 24 hr to an inoculum of 2 $\times$ 10^8 organisms.[17] Although lung involvement was not reported, moribund mice showed histopathological changes in their lymphoid organs, crypts of the small intestine, and their livers.[17] This model is also postulated to represent a toxic syndrome.

A preliminary report of disease in rhesus monkeys by low-dose infection indicated that animals developed fever and histopathological changes consistent with Legionnaires' disease.[18]

Subcutaneous chambers, similar to those used to study immunity to gonocci, have been implanted in mice and guinea pigs and used as another way of investigating anti-*Legionella* immunity.[19]

II. IN VIVO PROTECTION STUDIES IN ANIMALS

Only a few in vivo protection studies have been reported using animals as laboratory models of infection, and in none of these has the aerosol route of infection been employed. Elliott et al.[20] immunized Hartley guinea pigs subcutaneously with FCA containing heat-killed cells (101°C for 1 hr) or with an antigenic fraction obtained from a serogroup 1 isolate and designated F-1. A booster dose was given three weeks later and all animals were challenged intraperitoneally two weeks after the booster with 100 LD_{50} doses. To test the cross-protective capacity of the vaccines, eight groups of immunized guinea pigs were used, and each group was vaccinated with heat-killed cells (10^8) or with F-1 antigen (0.5 mg) prepared from a different serogroup, either 2, 3, or 4. All animals were challenged with serogroup 1 cells. It was found that protection induced by heat-killed cells was serogroup specific. When the F-1 antigen was used for vaccination, one of four animals receiving the F-1 antigen of serogroups 2 or of 3 also survived. Thus serogroup-specific protection was elicited by the F-1 antigen as well as a low level of cross-serogroup protection.

Wong and collaborators[21,22] have carried out protection studies in guinea pigs using two different antigen preparations. Both antigens were purified from agar-grown organisms disrupted with lysozyme and treated with deoxyribonuclease. Crude supernatant material chromatographed on diethylaminoethyl cellulose (DEAE) yielded one peak containing a serogroup-specific antigen, and a second peak containing an antigen which cross-reacted with different serogroups of *Legionella pneumophila*.[21] The specific antigens contained branched-chain fatty acids and were lipid-protein-carbohydrate complexes, whereas the cross-reacting antigens were primarily protein.[21] As these antigens were purified before the isolation of all of the presently recognized *Legionella*, and before the adoption of a uniform terminology, the serogroup-specific antigens were designated by these investigators as "serotypic" antigens. The serotypic antigen (serogroup-specific) prepared from a Knoxville 1 strain was used to vaccinate seven guinea pigs i.p. Four injections (125 μg protein per dose) one week apart were given, and pigs were challenged one week after the last injection with 2×10^6 organisms of a Philadelphia 1 strain. All immunized guinea pigs survived, while all controls succumbed, thus demonstrating intraserogroup protection.[21] Cross-serogroup protection was not tested. Mice immunized subcutaneously with three weekly injections of 10 or 150 μg of the Knoxville 1 serotypic antigen were also protected against i.p. challenge with organisms of the Philadelphia 1 strain.[21] In a similar experiment 100% protection was demonstrated by i.p. challenge with the Bellingham organism. Mice (ICR strain) vaccinated with the Knoxville antigen were also shown to exert an inhibitory effect on the growth of the Philadelphia 1 strain implanted in subcutaneous chambers.[21]

In further studies, Wong et al.[22] carried out checkerboard protection experiments using organisms of serogroups 1, 2, and 4 and showed that an antigen of the homologous serogroup ("serotype") to the challenge strain gave the best protection, but in agreement with Elliott et al.,[20] some cross-serogroup protection was also observed. Extraction of the antigen with chloroform-methanol reduced the cross-protective activity somewhat (from 40% or 50% protection to 20%), although the number of animals in each group was only 10.[22]

Wong et al.[21] have also carried out in vivo passive serum protection studies using 40 mg of goat IgG raised to the Knoxville 1 "serotypic" antigen. The serum was given i.p. and guinea pigs were challenged 4 hr later. Protection was shown to be dependent on the challenge dose, with a significant number of animals surviving. Further, marked differences in survival between animals receiving immune IgG and those receiving normal goat IgG were seen. The same immune IgG also protected mice against challenge with mucin-coated organisms. The challenge doses reported by this group are expressed as LD_{50} doses required to kill chick embryos, rather than as LD_{50} doses for the challenged animal. Therefore, comparison of the magnitude of protection reported by this laboratory with that reported by Elliott et al. cannot be easily done.

Hedlund et al[17] showed that AKR/J mice could be protected against death, but not illness, by i.p. vaccination with a soluble antigen extracted with 0.5 *N* NaOH from a Philadelphia 1 strain. Challenge was i.p. with another serogroup 1 organism. Goat antisera raised to formalinized organisms afforded passive protection to mice when challenge was one week after serum transfer.

Rolstad and Berdal[16] using a rat model, in which the LD_{50} was 6×10^8 CFU of a serogroup 1 strain (designated LD-8), have shown that active immunization with Philadelphia 1 heat-killed cells (100°C for 15 min), or passive immunization with murine immune serum, gave significant protection. Protection was found to be serogroup specific.

In both the mouse and the rat model, animals became ill within 4 to 6 hr and died within 24 hr from a large bacterial inoculum. Thus both models are interpreted as representing a toxic death, possibly due to endotoxin, and the observed protection is seen as a possible model for anti-toxic immunity.

Experiments by Arko et al.[19] showed that guinea pigs immunized with formalinized cells of the Knoxville 1 strain were more resistant to subcutaneous chamber infection with that strain. However, paradoxically, immunized ICR mice implanted with subcutaneous chambers had better growth of organisms in the chamber than nonimmune controls.

It is evident that much work still remains to be carried out using in vivo model systems to adequately assess the nature of the protective immune response. No protection studies have been published to date using an aerosol, an intratracheal, or an intranasal route of infection. From the data presently available it can be concluded that guinea pigs can be protected against intraperitoneal infection by whole cells, by antigenic extracts, and by immune serum. Further, the protection occurs among different strains of the same serogroup with some evidence of a lesser degree of cross-serogroup protection. No experiments measuring protection between *Legionella* species have been reported.

Experiments using other models, which are perhaps less physiologic than the intraperitoneally challenged guinea pig, yield, with only one exception,[19] results that are consistent with those obtained in the pig. Thus, killed vaccines and antibody are seen to be protective.

These data can be interpreted in two ways; either that humoral immunity is effective against *Legionella*, or, alternatively, that the model systems chosen are not reflective of the conditions of host defenses in humans. Examination of evidence bearing on the mechanisms of host defense which comes from other kinds of experiments will be considered to help evaluate the significance of the in vivo protection results.

III. IN VIVO STUDIES OF DELAYED HYPERSENSITIVITY

Only a few studies have been carried out to look for delayed-type hypersensitivity (DTH) to *Legionella* antigens in infected people or animals. A skin test for DTH would be advantageous, as it would allow screening of the population for epidemiologic evidence of prior exposure to determine the extent of clinical disease. Further, the demonstration of DTH is considered evidence for the existence of cellular immunity.

Plouffe and Baird[23] showed that peripheral blood lymphocytes of 17 patients recovering from Legionnaires' disease had increased thymidine incorporation upon stimulation with an antigen of *L. pneumophila*, as compared with subjects with no history of clinical Legionnaires' disease. Although mean stimulation indices showed significant differences in the two populations, in this and a follow-up study[24] some of the "normal" controls did show significant label incorporation, perhaps suggesting prior exposure with subclinical disease. The interpretation of these results with regard to the mechanism of immunity is not clear, as the nature of the responding cells was not determined. Peripheral blood lymphocytes contain both T and B cell populations, although in humans about 80% are T cells. The authors state, but data were not given, that the responding cells are T cells. Blast transformation of T cells by specific antigen has been interpreted as evidence for cell-mediated immunity in other infectious diseases. Thus, these preliminary studies using *Legionella* antigen may be cautiously interpreted as providing possible support for T cell sensitization in *Legionella* infection.

Wong and colleagues have used the guinea pig to evaluate possible skin test antigens. Their first report showed positive responses in Hartley guinea pigs sensitized to Knoxville 1 cells given in FCA and elicited four to seven weeks later with "serotypic" antigen.[25] Further studies have shown that serogroup-specific antigens ("serotypic" antigens) of *L. pneumophila*, as well as antigens which are cross-reactive for at least 4 serogroups of the species, can elicit DTH in guinea pigs sensitized by injection of heat-killed cells in FCA.[25] Both antigens elicited positive skin reactions in guinea pigs sensitized with heat-killed cells in complete adjuvant. Of particular importance was the demonstration that the cross-reacting antigen could elicit DTH in animals sensitized 2 to 3 weeks previously be exposure to live organisms contained in a subcutaneously implanted chamber. Similar reactions to the common antigens were also obtained in animals which had first been immunized with the extracted serotypic antigens, then challenged with live organisms, and finally skin-tested 3 to 4 weeks after the infection. These studies show that a serogroup-specific and a cross-reactive antigen can elicit delayed hypersensitivity in artificially sensitized animals (killed organisms plus FCA), and that under certain specialized conditions of exposure to live organisms, guinea pigs develop DTH to the common antigen preparation.[25]

Friedman et al.[26] have shown that spleen cells of guinea pigs given *Legionella* in FCA show macrophage migration inhibition when exposed to *Legionella* sonicates in vitro. Similarly immunized animals also give splenic blast cell transformation.[27]

IV. IN VITRO IMMUNITY STUDIES

An alternative to costly in vivo protection studies is the use of in vitro culture systems where the effect of immune serum and lymphokines can be investigated with respect to phagocytosis, ingestion, invasion, or intracellular growth in either phagocytic cells, primary cell cultures, or in established tissue culture cell lines.

A. Interaction of *Legionella* with Professional Phagocytes

Kishimoto et al.[28] were the first to show that the Washington strain of *Legionella* could infect alveolar cells of cynomolgus monkeys in vitro when the ratios of organisms to monkey cells was 100:1. Uptake and intracellular replication were evident within 3 hr and extensive by 24 hr. Electron micrographs showed macrophages with cytoplasm packed with bacterial cells. Although only a small percentage of *Legionella* were phagocytized, those that were ingested apparently grew very well.

Johnson et al.[29] studied uptake of *Legionella* by rat alveolar macrophages and by mouse peritoneal macrophages. They found that ^{3}H-labeled organisms were interiorized in the presence of immune serum. Poor uptake was found without immune serum, and opsonization was serogroup specific. The specificity of the immune serum was demonstrated by removal of opsonic activity upon absorption with a purified homologous serogroup antigen, but not a heterologous preparation. Fresh serum, which presumably contained complement, did not promote phagocytosis in the presence or absence of immune serum.

Kishimoto et al.[30] have also reported on the interaction of *L. pneumophila* with guinea pig peritoneal macrophages in vitro. Variables tested were ingestion and intracellular growth of virulent as compared with avirulent bacteria, the importance of immune serum, and the influence of previous in vivo infection on the behavior of the phagocytic cells. It was found that normal macrophages required immune serum to phagocytize *Legionella* whether virulent or avirulent, whereas macrophages from previously infected animals were more phagocytic. Virulent organisms multiplied intracellularly, whereas the avirulent strain did not, although no data was presented to show that equal numbers of both organisms were ingested. Immune serum, while promoting ingestion, did not inhibit intracellular replication. Interestingly, macrophages from immune animals supported intracellular bacterial growth, rather than exhibiting bacteriostasis. The results were obtained by visual observation using scanning and transmission electron microscopy, and are therefore only qualitative.

Horwitz and Silverstein[31] have shown that *L. pneumophila*, serogroup 1, will infect and grow in human monocytes purified from peripheral blood. Infection ratios of 1 organism to 100 mononuclear cells gave the best intracellular growth, with peak infection at 4 days after initiation of the culture. Monocytes were purified on a Ficoll-sodium diatrizoate solution and cultured in 15% autologous human serum which had no anti-*Legionella* titer by indirect immunofluorescence. Intracellular organisms were shown to be enclosed in ribosome-studded vesicles. Additional studies by these investigators showed that human polymorphonuclear leukocytes ingest *Legionella*, but interiorization only occurred efficiently in the presence of both antibody and complement.[32] Further, it was found that ingested organisms were poorly killed by the PMNs, although no evidence of intracellular multiplication was found.[33] Studies by Friedman et al.[34] provide a mechanism by which *Legionella* might resist the bactericidal mechanisms of the PMN, for they have reported that a Knoxville 1 strain can inhibit O_2 consumption and NADPH production in human cells, suggesting interference with the oxidative PMN killing pathway.

No evidence of in vitro bactericidal activity was found by Horwitz and Silverstein for fresh normal, human serum alone (without antibody), or with the addition of rabbit or human anti-*Legionella* antibody,[33] observations which suggest that human antibody plus complement are not lethal for this organism. In contrast, Arko et al.[19] found that fresh human serum was bactericidal for a Knoxville 1 isolate and a Bellingham 1 isolate. The bactericidal activity was heat-labile at 56°C for 30 min, suggesting a complement-mediated killing mechanism. As human serum deficient in C4 was not bactericidal, the results were interpreted to represent activation of the classical-complement pathway. Since no antibody determinations were made on the serum, its contribution to the observed killing could not be assessed. Somewhat paradoxically, mouse serum with antibodies to the Knoxville 1 strain was found to "block" the bactericidal effect of human complement on this strain.

The parameters of the interaction of *Legionella* with human monocytes have been investigated further by Horwitz and Silverstein. They have shown that human antibody plus complement promotes association of *Legionella* with monocytes in culture and mediates a modest amount of bacterial killing by those cells. However, the majority of the bacteria survive in spite of their antibody-complement coat and proceed to replicate intracellularly.[32] In another paper[35] these same investigators reported that human, peripheral blood monocytes incubated with lymphocytes and concanvalin A, or with supernatants from such cultures, inhibited the intracellular multiplication of *Legionella*. Activated monocytes ingested fewer bacteria, but also exerted bacteriostatic control over those which were interiorized.

B. Interaction of *Legionella* with Tissue Cultured Cells

Legionella have been shown to invade and grow intracellularly in tissue cultured cells which are not considered professional phagocytes. Thus, Wong et al.[36] have demonstrated that Philadelphia 1 and Knoxville 1 organisms will grow in cultured human embryonic lung fibroblasts. Daisy et al.[37] have cultured the Philadelphia 2 strain of *Legionella* in MRC-5, HeLa, Hep2, and McCoy cells. Intracellular growth was confirmed by transmission electron microscopy. *L. micdadei* (Pittsburgh Pneumonia Agent) has been shown to replicate intracellularly in Vero cells, HeLa cells, human foreskin cells, and chicken embryo cells.[38]

V. SUMMARY

Thus *Legionella* have been shown to enter and multiply in a variety of primary and tissue culture cell lines. Further, they have been shown to replicate in culture in alveolar cells of the cynomolgus monkey, in guinea pig peritoneal macrophages, and in human monocytes. They appear to be in ribosome-studded vesicles in human and guinea pig cells.[12,31]

The preponderance of evidence from histopathological studies of human and animal tissues, as well as in vitro experiments using phagocytes or cell lines, indicates that *Legionella* are facultative intracellular pathogens. The studies of Horwitz and Silverstein would seem to be the most relevant in this regard, as they have shown intracellular multiplication of *Legionella* in human monocytes, and inhibition of this intracellular growth in monocytes treated with supernatants of concanavalin A-stimulated cultures of lymphocytes and monocytes. These studies certainly demonstrate an in vitro correlate of cellular immunity. Whether the interaction of *Legionella* with the more mature human alveolar macrophage, the cell which presumably the organism would encounter if the natural route of infection were via aerosol, will be the same, remains to be determined. Key points which remain to definitively complete the evidence for cellular immunity in controlling *Legionella* replication, are (a) to show that T cells are the mediators of the activation of monocytes to macrophages with enhanced bacteriostatic activity, and (b) that T cells can transfer immunity in the guinea pig model. In studies of mitogenesis of peripheral blood lymphocytes obtained from patients and stimulated with *Legionella* antigens, it remains to be shown that the responding cells are T cells. In guinea pigs, definitive studies to show that sublethal infection gives delayed hypersensitivity are yet to be accomplished.

Considerable evidence also exists from studies with guinea pigs, rats and mice to support a protective role of antibody. Goat immune serum and its IgG fraction have been shown to passively protect guinea pigs against intraperitoneal infection. Second, heat-killed *Legionella* as well as *Legionella* antigens, have been shown to protect guinea pigs against i.p. challenge. It is a generally held tenet that nonliving cells, unless given with FCA, do not induce T cell immunity. Thus, the demonstration that heat-killed cells or purified antigens can protect in the absence of FCA suggests a protective role for antibody. Antibody also appears to be important for efficient phagocytosis of *Legionella* by macrophages and PMNs of both humans and guinea pigs, although it does not appear to mediate intracellular killing as assessed by in vitro bactericidal assays. A crucial unanswered point is whether nonliving vaccines and passive serum can protect pigs against aerosol challenge. Further, an evaluation of the importance of secretory IgA in host defense via aerosol challenge of the guinea pig has not been approached.

The validity of the animal studies which show a major role for antibody and also that killed vaccines are effective, is a crucial issue. Critical interpretation of the results hinges on how closely the animal models reflect the human condition.

Infection by *Salmonella typhi* provides a well-studied illustration of an intracellular pathogen whose only natural host is man. To achieve infection with this organism in mice, large numbers of bacteria (in excess of 1×10^7) must be injected, or organisms must be coated with mucin to increase virulence.[39] All evidence indicates that in the mouse, *S. typhi* grows extracellularly. Such a model, while useful for certain purposes, cannot be used to distinguish the relative importance of humoral and cellular immunity in host defense, as antibody is fully protective in this model. However, the importance of antibody in the human, where the organism presumably has an intracellular habitat, is less certain.[40]

Further, in murine models of *Salmonella* infection using *S. typhimurium*, which is a natural mouse pathogen, and is virulent for mice, the relative contributions of cellular and humoral immunity seem to vary with the mouse strain.[40] Thus, the limitations of interpretation of experimental models must be carefully considered.

The rat and the mouse models of *Legionella* infection, which require large doses of organisms to produce illness, and in which antibody is protective, seem to be analogous to the *S. typhi* mouse model. As the authors point out, these models seem to measure protection against a toxic death, a parameter which may have relevance to certain selected aspects of human Legionnaire's disease, but the exact applicability is yet to be defined.

Whether guinea pig intraperitoneal infection will prove to be a proper model for assessing host defenses to a disease which is believed to be initiated via aerosol remains to be seen.

Another major unanswered question concerns the degree of cross-protection afforded humans by previous infection with *Legionella* of a particular serogroup. The studies carried out in guinea pigs using killed vaccines or antigenic extracts as immunogens show excellent protection against intraperitoneal challenge with the homologous strain as well as against other isolates in the same serogroup, and also a low level of cross-serogroup protection. The degree of cross-protection afforded by similar immunogens to an aerosol challenge is unknown. In addition, the demonstration of a common "protein" antigen which elicits delayed hypersensitivity across serogroups could suggest that attenuated organisms might induce broader cross-reactivity. This common antigen might be altered or destroyed by heat-killing of cells.

The last point to be considered is the role of an exotoxin in *Legionella* pathogenesis and immunity. To date, toxic activity has only been looked for in a single serogroup 1 isolate. The prevalence of the toxin in various isolates within the serogroup, in organisms in different serogroups, and in the different *Legionella* species is unknown. Whether antibodies to the toxin will be protective, and whether they would cross-react with putative toxins from other isolates, remains to be determined.

In conclusion, while it appears clear that *Legionella* are facultative intracellular pathogens, our dissection of the relative contributions of cellular and humoral immunity to host defense in the human are only beginning.

ACKNOWLEDGMENT

Supported by grants AI 16804 and AI 16618 from the National Institutes of Health.

REFERENCES

1. **Mackaness, G. B.,** Resistance to intracellular infection, *J. Inf. Dis.*, 123, 439, 1971.
2. **Collins, F. M.,** Cellular antimicrobial immunity, *CRC Crit. Rev. Microbiol.*, 7, 27, 1979.
3. **Krahenbuhl, J. L., Ruskin, L., and Remington, J. S.,** The use of killed vaccine in immunization against an intracellular parasite: *Toxoplasma gondii*, *J. Immunol.*, 108, 425, 1972.
4. **Friedman, R. L., Iglewski, B. H., and Miller, R. D.,** Identification of a cytotoxin produced by *Legionella pneumophila*, *Infect. Immun.*, 29, 271, 1980.
5. **Hedlund, K. W. and Larson, R.,** The identification of a *Legionella pneumophila* toxin with *in vivo* lethality, in *Bacterial Vaccines*, Robbins, J., Hill, J., and Sadoff, J., Eds., Thieme-Stratton, New York, 1982, 279.
6. **Wong, K. H., Moss, C. W., Hochstein, D. H., Arko, R. J., and Schalla, W. O.,** Endotoxicity of the Legionnaires' disease bacterium, Int. Symp. Legionnaires' Disease, *Ann. Intern. Med.*, 90, 624, 1979.
7. **Youmans, G. P.,** *Tuberculosis*, W. B. Saunders, Philadelphia, 1979.
8. **Mackaness, G. B., Blanden, R. V., and Collins, F. M.,** The immunology of antituberculous immunity, *Am. Rev. Respir. Dis.*, 97, 337, 1968.
9. **Saravolatz, L. D., Burch, K. H., Fisher, E., Madhavan, T., Kiani, D., Neblett, T., and Quinn, E. L.,** The compromised host and Legionnaires' disease, *Ann. Intern. Med.*, 90, 533, 1979.
10. **Cutz, E., Thorner, P. S., Rao, P., Toma, S., Gold, R., and Gelfand, E. W.,** Disseminated *Legionella pneumophila* infection in an infant with severe combined immunodeficiency, *J. Pediat.*, 100, 760, 1982.

11. **Chandler, F. W., McDade, J. E., Hicklin, M. D., Blackmon, J. A., Thomason, B. M., and Ewing, E. P.,** Pathologic findings in guinea pigs inoculated intraperitoneally with the Legionnaires' disease bacterium, *Ann. Intern. Med.*, 90, 671, 1979.
12. **Katz, S. M. and Hashemi, S.,** Electron microscopic examination of the inflammatory response to *Legionella pneumophila* in guinea pigs, *Lab. Invest.*, 46, 24, 1982.
13. **Winn, W. C., Jr., Davis, G. S., Gump, D. W., Craighead, J. E., and Beaty, H. N.,** Legionnaires' penumonia after intratracheal inoculation of guinea pigs and rats, *Lab. Invest.*, 47, 568, 1982.
14. **Davis, G. S., Winn, W. C., Jr., Gump, D. W., Craighead, J. E., and Beaty, H. N.,** Legionnaires' penumonia after aerosol exposure of guinea pigs and rats, *Am. Rev. Respir. Dis.*, 126, 1050, 1982.
15. **Berendt, R. F., Young, H. W., Allen, R. G., and Knutsen, G. L.,** Dose-response of guinea pigs experimentally infected with aerosols of *Legionella pneumophila*, *J. Infect. Dis.*, 141, 186, 1980.
16. **Rolstad, B. and Berdal, B. P.,** Immune defenses against *Legionella pneumophila* in rats, *Infect. Immun.*, 32, 805, 1981.
17. **Hedlund, K. W., McGann, V. G., Copeland, D. S., Little, S. F., and Allen, R. G.,** Immunologic protection against the Legionnaires' disease bacterium in the AKR/J mouse, *Ann. Intern. Med.*, 90, 676, 1979.
18. **Baskerville, A., Broster, M., Fitzgeorge, R. B., Hambleton, P., and Dennis, P. J.,** Experimental transmission of Legionnaires' disease by exposure to aerosols of *Legionella pneumophila*, *Lancet*, 26, 1389, 1981.
19. **Arko, R. J., Wong, K. H., and Feeley, J. C.,** Immunologic factors affecting the *in vivo* and *in vitro* survival of the Legionnaires' disease bacterium, *Ann. Intern. Med.*, 90, 680, 1979.
20. **Elliott, J. A., Johnson, W., and Helms, C. M.,** Ultrastructural localization and protective activity of a high-molecular-weight antigen isolated from *Legionella pneumophila*, *Infect. Immun.*, 31, 822, 1981.
21. **Wong, K. H., Schalla, W. O., Arko, R. J., Bullard, J. C., and Feeley, J. C.,** Immunochemical, serologic and immunologic properties of major antigens isolated from the Legionnaires' disease bacterium, *Ann. Intern. Med.*, 90, 634, 1979.
22. **Wong, K. H., Schalla, W. O., Wong, M. C., McMaster, P. R. B., Feeley, J. C., and Arko, R. J.,** Biologic activities of antigens from *Legionella pneumophila*, in *Bacterial Vaccines*, Robbins, J., Hill, J., and Sadoff, J., Eds., Thieme-Stratton, New York, 1982, 434.
23. **Plouffe, J. F. and Baird, I. M.,** Lymphocyte transformation to *Legionella pneumophila*, *J. Clin. Lab. Immunol.*, 5, 149, 1981.
24. **Plouffe, J. F. and Baird, I. M.,** Lymphocyte blastogenic responses to *L. pneumophila*, *J. Clin. Lab. Immunol.*, 7, 43, 1982.
25. **Wong, K. H., McMaster, P. R. B., Feeley, J. C., Arko, R. J., Schalla, W. O., and Chandler, F. W.,** Detection of hypersensitivity to *Legionella pneumophila* in guinea pigs by skin test, *Curr. Microbiol.*, 4, 105, 1980.
26. **Widen, R., Klein, T., Lee, I., and Friedman, H.,** An agar droplet assay for cell-mediated immunity to *Legionella* antigen, *J. Clin. Microbiol.*, 17, in press, 1983.
27. **Widen, R., Lee, I., Klein, T., and Friedman, H.,** Blastogenic responsiveness of spleen cells of guinea pigs sensitized to *Legionella* antigen, *Proc. Soc. Exp. Biol. Med.*, 1983, in press.
28. **Kishimoto, R. A., Kastello, M. D., White, J. D., Shirey, F. G., McGann, V. G., Larson, E. W., and Hedlund, K. W.,** In vitro interaction between normal cynomolgus monkey alveolar macrophages and Legionnaires' disease bacteria, *Infect. Immun.*, 25, 761, 1979.
29. **Johnson, W., Pesanti, E., and Elliott, J.,** Serospecificity and opsonic activity of antisera to *Legionella pneumophila*, *Infect. Immun.*, 26, 698, 1979.
30. **Kishimoto, R. A., White, J. D., Shirey, F. G., McGann, V. G., Berendt, R. F., Larson, E. W., and Hedlund, K. W.,** In vitro response of guinea pig peritoneal macrophages to *Legionella pneumophila*, *Infect. Immun.*, 31, 1209, 1981.
31. **Horwitz, M. A. and Silverstein, S. C.,** Legionnaires' disease bacterium (*Legionella pneumophila*) multiplies intracellularly in human monocytes, *J. Clin. Invest.*, 66, 441, 1980.
32. **Horwitz, M. A. and Silverstein, S. C.,** Interaction of the Legionnaires' disease bacterium (*Legionella pneumophila*) with human phagocytes. I, *J. Exp. Med.*, 153, 398, 1981.
33. **Horwitz, M. A. and Silverstein, S. C.,** Interaction of the Legionnaires' disease bacterium (*Legionella pneumophila*) with human phagocytes. II, *J. Exp. Med.*, 153, 386, 1981.
34. **Friedman, R. L., Lochner, J. E., Bigley, R. H., and Iglewski, B. H.,** The effects of *Legionella pneumophila* toxin on oxidative processes and bacterial killing of human polymorphonuclear leukocytes, *J. Infect. Dis.*, 146, 328, 1982.
35. **Horwitz, M. A. and Silverstein, S. C.,** Activated human monocytes inhibit the intracellular multiplication of Legionnaires' disease bacteria, *J. Exp. Med.*, 154, 1618, 1981.
36. **Wong, M. C., Ewing, E. P., Jr., Callaway, C. S., and Peacock, W. L., Jr.,** Intracellular multiplication of *Legionella pneumophila* in cultured human embryonic lung fibroblasts, *Infect. Immun.*, 28, 1014, 1980.

37. **Daisy, J. A., Benson, C. E., McKitrick, J., and Friedman, H. M.,** Intracellular replication of *Legionella pneumophila*, *J. Infect. Dis.*, 143, 460, 1981.
38. **Rinaldo, C. R., Jr., Pasculle, A. W., Myerowitz, R. L., Gress, F. M., and Dowling, J. N.,** Growth of the Pittsburgh pneumonia agent in animal cell cultures, *Infect. Immun.*, 33, 939, 1981.
39. **Landy, M., Johnson, A. G., Webster, M. E., and Sagin, J. F.,** Studies on the O antigen of *Salmonella typhosa*, *J. Immunol.*, 74, 466, 1955.
40. **Eisenstein, T. K. and Sultzer, B. M.,** Immunity to *Salmonella* infection, in *Host Defense to Intracellular Pathogens*, Eisenstein, T. K., Actor, P., and Friedman, H., Eds., Plenum Press, New York, 1983, 261.

Part IV: The New Microbe: Commentary

Chapter 11

THE DISCOVERY OF HEPATITIS B VIRUS: ONE PROTOTYPE FOR DISCOVERY OF NEW MICROBES

Baruch S. Blumberg

TABLE OF CONTENTS

I. INTRODUCTION

The etiologic agent of Legionnaires' disease was discovered as the result of a large, intensive, and coordinated investigation of a catastrophic epidemic; the dramatic presentation of the disease resulted in the eventual identification of an infectious agent. The editor has assigned me the task of describing the discovery of the hepatitis B virus (also a Philadelphia event), to compare it to the discovery of *Legionella* and to comment on the nature of discovery of infectious agents. Before proceeding to this assignment, I would like to provide some comments on the Philadelphia *Legionella* outbreak.

Nineteen hundred and seventy six was, in most respects, a wonderful year in Philadelphia. The celebration of the bicentennial of the founding of the United States was lively and extended; a year-long series of fairs, plays, fireworks, huge block parties, parades, bands, fife & drum corps, and parties. This all culminated in a massive parade on the Fourth of July, beautiful and brilliant, and the city settled back to enjoy the remaining months of the summer and to welcome additional visitors. The tragic Legionnaires' disease epidemic changed all this; after the national publicity it generated, visitors did not come to Philadelphia in the numbers expected, and the citizens of the city participated in the final months of ceremonies by themselves.

Our home is only a few blocks from Independence Hall and the other national monuments. We often view the many parades that wind by Independence Hall, and my son and I saw the Legionnaires' parade the evening before the day the first cases of what became known as Legionnaires' disease were reported. I remarked to my wife at dinner that evening that some of the participants in the parade appeared to be in poor general physical condition. Early analysis of the data showed that it was those with underlying illness who had the worst outcomes of *Legionella* infection — a corollary of the intuitive observations that healthy people fare better with disease than those less fit.

During the extensive investigation by the Center for Disease Control of the epidemic, I was called to Harrisburg by the then Secretary of Health, Dr. Leonard Bachman, as a consultant. I was impressed by the large number and the dedication of the CDC officials assigned to this problem. It occurred to me later, when the solution had been reached, that although there had been previous epidemics, the etiologic agent was not found until the 1976 Philadelphia epidemic. This could be attributed, in part, to the attention focused and sustained on the outbreak by the Pennsylvania and Philadelphia authorities, which eventuated in the intensive investigation and the skillful detective work of the CDC scientists.

II. THE DISCOVERY OF THE HEPATITIS B VIRUS

Legionella was discovered as the result of a search for the etiology of what was believed to be a "new" disease. Hepatitis B virus (HBV) was discovered in an apparently normal person, and the research was directed towards determining its disease association. As a consequence, much of the research on etiology was done on the sera of people who were not apparently ill from hepatitis infection.

The investigation began as a consequence of the interest in our laboratory in inherited variation in human serum. It was hypothesized that if polymorphic antigenic variants existed in serum proteins, as they do on red blood cell antigens, then some patients who received large numbers of transfusions might develop antibodies against serum proteins they had not inherited. This hypothesis was tested using, as a putative antiserum, serum obtained from patients who had received many transfusions. These antisera were tested against a panel of normal sera that might contain the unknown antigens that generated the antibody. Testing was done using the technique of double diffusion in agar gel, which is a highly specific, but insensitive, immunologic technique. We soon found antibodies in the blood of a person with an unusual form of aplastic anemia who had received many transfusions. This antiserum (and several others which were subsequently found) defined a series of inherited specificities on the low density lipoproteins, and we spent several years trying to understand this "new" antigenic polymorphism.

We also continued to search for additional antisera that might define other antigenic systems present in human blood, on the notion that if this approach was successful in revealing one system, it might also, in due course, uncover others. In 1963, a new kind of precipitin band was seen between the "antiserum" from a frequently transfused hemophilia patient from New York, and the serum of an Australian aborigine that was included in the test panel. The Australian sample had been included as part of our general search strategy because it was known that the genes determining polymorphisms often vary in frequency from population to population. Since we were looking for an unknown polymorphism, we included sera from many populations in our panel in order to increase the probability of screening populations with high frequencies of the putative polymorphic gene. Hence, our panel included, in addition to sera from local populations, samples from foreign populations including Africans, Asians, Oceanians and, in this case, Australian aborigines.

We were now faced with the problem of determining the nature of the "Australia antigen" as we termed it. In order to investigate this question, we had to develop a series of hypotheses whose testing would shed light on this phenomenon. In order to generate hypotheses, it was necessary to have additional data on the distribution and character of the precipitin band. We soon learned that the Australia antigen was very stable. Serum that had been stored frozen for many years still retained detectable antigen and we were able to use a large number of sera from the blood collection of the Division of Clinical Research at the Institute for Cancer Research. From these surveys, which eventually included the testing of thousands of sera, we learned that the Australia antigen was rare in the U.S.; using the immunodiffusion technique, it occurred in only about one of two thousand of normal people tested. It was, however, common in various parts of Africa, Asia, and Oceania.

We also tested sera from patients from the U.S. and found that the antigen was common in leukemia. Based on these population and disease observations, we made a series of hypotheses. One of these was that the Australia antigen causes leukemia, the second was that the leukemia caused the Australia antigen, and the third that there was a common factor that caused individuals to have both leukemia and persistent Australia antigen. A corollary of the last hypothesis is that a high frequency of Australia antigen would be expected in groups of patients with a high likelihood of developing leukemia. This hypothesis was tested in patients with Down's syndrome who are known to have a higher risk of developing leukemia. The prediction was fulfilled; about one third of Down's syndrome patients cared for in large institutions were found to have the Australia antigen, a frequency much higher than other mentally retarded children in the same institution or in people in the general community. In addition to providing support for the hypothesis, this study revealed a group of individuals on whom observations could be made. (Down's syndrome patients were not used for experimental purposes; that is, they were not injected or immunized with experimental materials. Careful clinical observations were made on these patients as part of their regular care procedures.)

We soon established that if an individual was "positive" for Australia antigen on first testing, then on subsequent testing he or she was also positive; those found to be negative for the Australia antigen when first tested, were consistently negative. Early in 1966, we found an exception to this rule. One of our patients, a young man named James Bair who had originally been negative, was found to be positive on a subsequent test. He was admitted to our Clinical Research unit for observation. Since he now had a protein he did not previously have, and since proteins are often produced in the liver, we performed a series of liver function tests. These revealed that between the time James Bair had been negative for the Australia antigen and the time he became positive for it, he had developed a form of anicteric hepatitis. This observation led us to test the hypothesis that Australia antigen was on or a part of the putative virus that caused the disease.

In retrospect, there had, in fact, been other observations which occurred prior to the incident in our Down's syndrome patient that could have generated the Australia antigen-hepatitis hypothesis. For example, we had sampled a large number of sera from what we believed to be a "normal" population of a community in eastern Georgia. One of these was found to have Australia antigen. On checking back with our physician-colleague who had supplied these specimens to us, we learned that the positive individual rather than having been a member of the general population was a patient and that he had hepatitis. In another instance, we had tested a large number of employees at the National Institutes of Health and again found one person who had Australia antigen. On further inquiry, we learned from her physician that she had received a transfusion sometime prior to the testing. It's interesting that although these observations were noted and the possible connection with hepatitis considered, we apparently were very much more affected by the observation in our own patient and it was this that accelerated the testing of the Australia antigen-hepatitis hypothesis.

About this time, we made antiserum and antigen available to other investigators, and confirmation of the hepatitis connection came from many sources. Meanwhile, we tested the hypothesis that the Australia antigen was, in fact, part of hepatitis virus. One form of the virus particle was seen in the electron microscope, transmission under certain circumstances to infant African Green monkeys was demonstrated, transmission by blood tranfusion was observed, and certain other characteristics indicated that we had, in fact, identified a virus. It became obvious immediately that the test for Australia antigen could be used to identify hepatitis blood donors who were occult carriers of the hepatitis virus, particularly when more sensitive tests such as the radioimmunoassay were introduced. In a very short time, requirements were established by the blood transfusion service of the American Red Cross, the Association of Blood Banks and the Federal government that required the testing of all donor blood for the hepatitis B virus prior to its use for transfusion. Similar regulations were soon enforced in other parts of the world.

It has been estimated that the saving in medical costs and time lost from work as a consequence of the use of this test is now in the order of billions of dollars — a figure far exceeding any money spent on all the research on hepatitis.

Thus, the hepatitis virus was discovered in an unconventional manner by the use of techniques (the immunodiffusion method) not ordinarily employed for the discovery of viruses. The initial identifications of the virus were in apparently normal people and the investigation consisted of finding the disease associated with the Australia antigen. This is in interesting contrast to the discovery of the *Legionella* organism which was discovered as a result of isolating the microbe during an intensive investigation of a disease outbreak.

ACKNOWLEDGMENT

This work was supported by USPHS grants CA-06551, RR-05539 and CA-06927 from the National Institutes of Health and by an appropriation from the Commonwealth of Pennsylvania.

Chapter 12

MICROBIAL PARALLELS

Burton J. Landau and Amedeo Bondi

Since 1976 a remarkable amount of data has been acquired concerning the properties of *Legionella pneumophila* and the pathogenesis of Legionnaires' disease.[1] *L. pneumophila* is a nutritionally fastidious bacterium that possesses many characteristics exhibited by other gram negative microorganisms. Parallels exist between the pathogenesis of Legionnaires' disease and several clinical respiratory syndromes.[2,3] These diseases and their respective etiologic agents are listed in Table 1. This brief review will compare features of pathogenesis of these diseases with those of Legionnaires' disease.

Like each of the microbial agents listed in Table 1, *L. pneumophila* has the capacity to cause human respiratory disease following inhalation of infected droplet nuclei. Nontuberculous mycobacterial infection can also be acquired by the oropharyngeal route.[4] The origin of infective aerosols, and the mode of transmission varies considerably from one disease to another. *L. pneumophila*, for example, has often been found in ponds, lakes, rivers, etc., while the saprobic phase of both *H. capsulatum* and *C. immitis* is associated with a soil habitat. *C. burnetii* and *C. psittaci* are commonly found in discharges and other body fluids produced by infected animals or in dust contaminated by these organisms. For *C. burnetti, H. capsulatum*, and *C. immitis*, person to person transmission of disease seldom occurs, and it has not conclusively been demonstrated for either *L. pneumophila* or nontuberculous mycobacteria. This suggests that the number of virulent organisms needed to initiate these diseases exceeds that likely to be found in aerosols derived from respiratory discharges of infected persons. This concept is supported by the observation that tuberculosis in man can occur following inhalation of as few as ten organisms or less.[5] Moreover, serious disease caused by these agents, whose habitat is soil or water, occurs more frequently in immunocompromised hosts or in those individuals whose genetic background renders them more susceptible to a given disease, e.g., coccidioidomycosis in the American Indian.[6] Of those organisms listed in Table 1, only tuberculosis, plague pneumonia, influenza, and mycoplasma pneumonia are frequently transmitted from person to person.

Following inhalation, each of the agents listed in Table 2, except for influenza virus and *M. pneumoniae*, frequently gain direct access to the lower respiratory tract. Such access is facilitated in individuals who smoke or who have inhaled other noxious substances, since ciliary activity in these people is likely to be impaired.[5] Most of these organisms invade the lower respiratory tract, but to date, anatomically specialized surface structures for specific attachment to susceptible cells has not been demonstrated. Even *C. immitis* and *H. capsulatum*, which possess specific tissue forms yet appear to lack structures for attachment. Flagella and fimbriae as revealed by electron microscopic studies of laboratory strains of *L. pneumophila* have not been linked to the tissue trophisms of legionellosis.[8,9] Surface ligands of chlamydia do apparently mediate interaction with host cells or "nonprofessional" phagocytes.[10,11]

L. pneumophila and most of the other microorganisms listed in Table 2 are phagocytosed in the lower respiratory tract by alveolar macrophages. In the absence of specific immunity, these phagocytes provide a haven for initial intracellular replication of the invading pathogen. Except for *C. immitis* whose replicative cycle is unique, the capacity of organisms to multiply within the macrophages of the lower respiratory tract probably represents an essential step in development of a microbial mass sufficiently large for subsequent dissemination throughout the body. During the initial invasion of the body, arthrospores of *C. immitis* are engulfed

Table 1
ETIOLOGIC AGENTS CAUSING RESPIRATORY DISEASE MANIFESTATIONS SIMILAR TO LEGIONNAIRES' DISEASE

Etiologic agent	Respiratory disease
Mycobacterium tuberculosis	Tuberculosis
Mycobacterium kansasii	Nontubercular pneumonia
Mycobacterium avium-intracellulare complex	Nontubercular pneumonia
Histoplasma capsulatum	Histoplasmosis
Coccidioides immitis	Coccidioidomycosis
Mycoplasma pneumoniae	Mycoplasma pneumonia
Chlamydia psittaci	Psittacosis
Coxiella burnetii	Q fever
Yersinia pestis	Plague pneumonia
Influenza virus	Influenza

by macrophages, and then converted to spherules, which then form endospores. It is the endospore that serves as an infectious center in various portions of the body.[12]

Studies have examined the mechanism(s) by which an obligate or facultative intracellular parasite escapes initial killing by phagocytes. Experiments done in vitro, show that a cytotoxin produced by *L. pneumophila* causes inhibition of oxygen-dependent activities in polymorphonuclear leukocytes following phagocytosis.[13,14] The inhibition results in a decrease in the ability of polymorphonuclear leukocytes to kill phagocytosed bacteria, but does not inhibit phagocytosis. These data support the suggestion of Horwitz and Silverstein[15] that polymorphonuclear leukocytes are not a major host defense against *L. pneumophila*. Other studies involving growth of *L. pneumophila* in guinea pig macrophages cultured in vitro have revealed that virulent strains survive phagocytosis and are able to multiply intracellularly[16] and reside in structures that morphologically resemble rough endoplasmic reticulum.[17] In these experiments, and in others involving macrophages and mycobacteria[18] and *C. immitis*,[19] inhibition of phagocytic killing is postulated to involve failure of phagolysosomal fusion. Eisenberg et al.[20] also suggest that persistence of ultraviolet-irradiated *C. psittaci* elementary bodies in mouse resident peritoneal macrophages is due to inhibition of phagolysosomal fusion, and the capacity of *C. psittaci* elementary bodies to inhibit phagolysosomal fusion in continuous cell cultures has been shown to be due to a membrane protein.[21] By contrast, persistence of *C. burnetii* in murine macrophage-like cultures occurs despite phagolysosomal fusion,[22] and survival of that microbe was attributed to its resistance to lysosomal hydrolases.

Resolution of infection by *L. pneumophila*, *M. tuberculosis*, *H. capsulatum*, and *C. immitis* is predominantly a function of cell-mediated immune mechanisms.[27-29] Humoral antibody may also have a role in protective immunity developed in response to these agents, but it is likely to be limited in importance. For example, while humoral antibody aids in the phagocytosis of *C. immitis* by polymorphonuclear leukocytes[12] and increases binding of *L. pneumophila* to monocytes,[30] the number of organisms either killed or inhibited from multiplying is relatively small. In addition, cell-mediated responses appear to be important in determining the outcome of mycoplasmal pneumonia[24] and psittacosis.[37] In influenza, cytotoxic T-lymphocytes have been postulated to be involved in resolving the acute infection.[38] Humoral antibody directed against the attachment proteins of *M. pneumoniae*[31,32] and influenza virus[33,34] do inhibit spread of infection and confer at least partial protection to reinfection. Antibody to *C. psittaci* inhibits its penetration of nonprofessional phagocytes[35] and enhances its phagocytosis by macrophages.[36] In the latter instance however, intracellular killing of chlamydia occurs by phagolysosomal fusion.

Table 2
ACTIVITIES OF SPECIFIC AGENTS WHICH CONTRIBUTE TO THEIR CAPACITY TO CAUSE DISEASE

Activity	Occurrence of activity by [a]										
	L. pneumophila	*M. tuberculosis*	*M. kansasii*	*M. avium-intracellulare*	*H. capsulatum*	*C. immitis*	*M. pneumoniae*	*C. psittaci*	*C. burnetii*	*Y. pestis*	Influenza virus
Dimorphism	–	–	–	–	+	+	–	+	–	–	N/A
Specialized structure(s) for attachment to host cells	–	–	–	–	–	–	+	–	–	–	+
Phagocytosis by alveolar macrophages	+	+	+/–	+/–	+	+	+	+	+/–	+	+
Obligate or facultative intracellular parasite	–	+	+	+	+	N/A	–	+	+	+	+
Hematogeneous or lymphogeneous spread throughout body	+	+	+	–	+	+	+	+	+	+	(–)

[a] + = Occurrence of specific activity; – = absence of specific activity; +/ = likely to occur but not proved; N/A = not applicable; (–) = may occur but not essential for disease production.

Parallels of infectious diseases often clarify and group properties of illnesses and microbes. The pathogenesis of Legionnaires' disease appears to be similar in many respects to that seen in pulmonary infections caused by other opportunistic facultative intracellular parasites. However, as critical events become better understood in pulmonary disease caused by *L. pneumophila* and other ubiquitous agents, perhaps distinctions will be more apparent. In any event, the increase in the reported incidence of pneumonia by *L. pneumophila* and the recognition that a substantial number of *Legionella* species exist suggest that control of this agent will continue to be difficult. Further studies are needed to determine more definitively the variables which affect the outcome of the host-parasite interaction.

REFERENCES

1. **Meyer, R. D.,** Legionella infections: a review of five years of research, *Rev. Infect. Dis.*, 5, 258, 1983.
2. **Band, J. D. and Fraser, D. W.,** Legionellosis (Legionnaires' disease and Pontiac fever), in *International Textbook of Medicine*, Vol. II, *Medical Microbiology and Infectious Diseases*, Braude, A. I., Ed., W. B. Saunders, Philadelphia, 1981, 958.
3. **Helms, C. M., Viner, J. P., Sturm, R. H., Renner, E. D., and Johnson, W.,** Comparative features of pneumococcal, mycoplasmal, and Legionnaires' disease pneumonias, *Ann. Intern. Med.*, 90, 543, 1979.
4. **Wolinsky, E.,** Nontuberculous mycobacteria and associated diseases, *Am. Rev. Respir. Dis.*, 119, 107, 1979.
5. **Mims, C. A.,** *The Pathogenesis of Infectious Diseases*, 2nd ed., Academic Press, New York, 1982.
6. **Sievers, M. L.,** Disseminated coccidioidomycosis among southwestern American Indians, *Am. Rev. Respir. Dis.*, 109, 602, 1974.
7. **Kahane, I., Banai, M., Razin, S., and Feldner, J.,** Attachment of mycoplasmas to host cell membranes, *Rev. Infect. Dis.*, 4(Suppl.), S185, 1982.
8. **Rodgers, F. G., Greaves, P. W., Macrae, A. D., and Lewis, M. J.,** Electron microscopic evidence of flagella and pili on *Legionella pneumophila*, *J. Clin. Pathol.*, 33, 1184, 1980.
9. **Chandler, F. W., Roth, I. L., Callaway, C. S., Bump, J. L., Thomason, B. M., and Weaver, R. E.,** Flagella on Legionnaires' disease bacteria. Ultrastructural observations, *Ann. Intern. Med.*, 93, 711, 1980.
10. **Byrne, G. I. and Moulder, J. W.,** Parasite specified phagocytosis of *Chlamydia psittaci* and *Chlamydia trachomatis* and L and HeLa cells, *Infect. Immun.*, 19, 598, 1978.
11. **Levy, N. and Moulder, J. W.,** Attachment of cell walls of *Chlamydia psittaci* to mouse fibroblasts (L cells), *Infect. Immun.*, 37, 1059, 1982.
12. **Drutz, D. J. and Huppert, M.,** Coccidioidomycosis: factors affecting the host-parasite interaction, *J. Infect. Dis.*, 147, 372, 1983.
13. **Friedman, R. L., Iglewski, B. H., and Miller, R. D.,** Identification of a cytotoxin produced by *Legionella pneumophila*, *Infect. Immun.*, 29, 271, 1980.
14. **Friedman, R. L., Lochner, J. E., Bigley, R. H., and Iglewski, B. H.,** The effects of *Legionella pneumophila* toxin on oxidative processes and bacterial killing of human polymorphonuclear leukocytes, *J. Infect. Dis.*, 146, 328, 1982.
15. **Horwitz, M. A. and Silverstein, S. C.,** Interaction of the Legionnaires' disease bacterium (*Legionella pneumophila*) with human phagocytes. I. *L. pneumophila* resists killing by polymorphonuclear leukocytes, antibody and complement, *J. Exp. Med.*, 153, 386, 1981.
16. **Kishimoto, R. A., White, J. D., Shirey, F. G., McGann, V. G., Berendt, R. F., Larson, E. W., and Hedlund, K. W.,** In vitro responses of guinea pig peritoneal macrophages to *Legionella pneumophila*, *Infect. Immun.*, 31, 1209, 1981.
17. **Katz, S. M. and Hashemi, S.,** Electron microscopic examination of inflammatory response to *Legionella pneumophila* in guinea pigs, *Lab. Invest.*, 46, 24, 1982.
18. **Goren, M. B., Hart, P. D., Young, M. R., and Armstrong, J. A.,** Prevention of phagolysosome fusion in cultured macrophages by sulfatides of *Mycobacterium tuberculosis*, *Proc. Natl. Acad. Sci. U.S.A.*, 73, 2510, 1976.
19. **Beaman, L. and Holmberg, C. A.,** In vitro response of alveolar macrophages to infection with *Coccidioides immitis*, *Infect. Immun.*, 28, 594, 1980.
20. **Eisenberg, L. G., Wyrick, P. B., Davis, C. H., and Rumpp, J. W.,** *Chlamydia psittaci* elementary body envelopes: ingestion and inhibition of phagolysosome fusion, *Infect. Immun.*, 40, 741, 1983.

21. **Zeichner, S. L.,** Isolation and characterization of macrophage phagosomes containing infectious and heat-inactivated *Chlamydia psittaci:* two phagosomes with different intracellular behaviors, *Infect. Immun.*, 40, 956, 1983.
22. **Akporiaye, E. T., Rowatt, J. D., Aragon, A. A., and Baca, O.,** Lysosomal response of a murine macrophage-like cell line persistently infected with *Coxiella burnetii, Infect. Immun.*, 40, 1155, 1983.
23. **Cassell, G. H. and Cole, B. C.,** Mycoplasmas as agents of human disease, *N. Engl. J. Med.*, 304, 80, 1981.
24. **Stanbridge, E. J.,** Mycoplasma-lymphocyte interactions and their possible role in immunopathologic manifestations of mycoplasma disease, *Rev. Infect. Dis.*, 4(Suppl.), S219, 1982.
25. **Mackowiak, P. A.,** Microbial synergism in human infections (first of two parts), *N. Engl. J. Med.*, 298, 21, 1978.
26. **Drutz, D. J. and Mills, J.,** Immunity and infection, in *Basic and Clinical Immunology*, 4th ed., Stites, D., Stabo, J. D., Fudenberg, H. N., and Wells, J. V., Eds., Lange Medical Publications, Los Altos, Calif., 1982, chap. 13.
27. **Horwitz, M. A. and Silverstein, S. C.,** Activated human monocytes inhibit the intracellular multiplication of Legionnaires' disease bacterium, *J. Exp. Med.*, 154, 1618, 1981.
28. **Domer, J. E. and Carrow, E. W.,** Immunity to fungal infections, in *Host Defenses to Intracellular Pathogens*, Eisenstein, T. K., Actor, P., and Friedman, H., Eds., Plenum Press, New York, 1983, 383.
29. **Collins, F. M.,** Cellular mechanisms of anti-mycobacterial immunity. In: *Host Defenses to Intracellular Pathogens*, Eisenstein, T., Actor, P., and Friedman, H., Eds., Plenum Press, New York, 1983, 157.
30. **Horwitz, M. A. and Silverstein, S. A.,** Interactions of Legionnaires' disease bacterium *Legionella pneumophilia)* with human phagocytes. II. Antibody promotes binding of *L. pneumophila* to monocytes but does not inhibit intracellular multiplication, *J. Exp. Med.*, 153, 398, 1981.
31. **Krause, D. C. and Baseman, J. H.,** Inhibition of *Mycoplasma pneumoniae* hemadsorption and adherance to respiratory epithelium by antibodies to a membrane protein, *Infect. Immun.*, 39, 1180, 1983.
32. **Hu, P. C., Huang, C-H., Collier, A. M., and Clyde, W. A., Jr.,** Demonstration of antibodies to *Mycoplasma pneumoniae* attachment protein in human sera and respiratory secretions, *Infect. Immun.*, 41, 437, 1983.
33. **Potter, C. W. and Oxford, J. S.,** Determinants of immunity to influenza infections in man, *Br. Med. Bull.*, 35, 69, 1979.
34. **Kilbourne, E. D., Butler, W. T., and Rossen, R. D.,** Specific immunity to influenza — Summary of influenza workshop. III, *J. Infect. Dis.*, 127, 220, 1973.
35. **Byrne, G. I. and Moulder, J. M.,** Parasite-specified phagocytosis of *Chlamydia psittaci* and *Chlamydia trachomatis* by L and HeLa cells, *Infect. Immun.*, 19, 598, 1978.
36. **Wyrick, P. B., Brownridge, E. A., and Ivins, B. E.,** Interaction of *Chlamydia psittaci* with mouse peritoneal macrophages, *Infect. Immun.*, 19, 1061, 1978.
37. **Lammert, J. K.,** Cytotoxic cells inducted after *Chlamydia psittaci* infection in mice, *Infect. Immun.*, 35, 1011, 1982.
38. **McMichael, A. J., Gotch, F. M., Noble, G. R., and Beare, P. A. S.,** Cytotoxic T-cell immunity to influenza, *N. Engl. J. Med.*, 309, 13, 1983.

Appendix

CHRONICLE, PHILADELPHIA, 1976

Donald Drake

The date was Monday, August 2, 1976.

Dr. Robert Craven, a pleasant, 32-year-old officer in the Epidemiologic Intelligence Service (EIS), was working in the swine flu monitoring section when the phone call came in to the Centers for Disease Control in Atlanta. It was from Dr. Sidney Franklin of the Veterans' Administration Clinic in Philadelphia.

Betraying none of the excitement or concern that he felt, Dr. Franklin explained that he had a patient with a high fever, joint pain, muscle tenderness, and general discomfort. Dr. Franklin conceded that such common symptoms were not particularly alarming; the reason he was calling was that his patient had attended an American Legion convention at which, according to the clinic grapevine, 26 other people had developed similar symptoms and four of them had died.

There was silence on the long-distance wire. Then Dr. Craven spoke, equally calmly.

"That's strange," he said. "We've heard absolutely nothing." Dr. Craven asked Dr. Franklin to get more information and call him back. Within the hour, Dr. Franklin was on the phone again, but this time he wasn't calm.

"How does 11 grab you?" He almost yelled it over the phone to Dr. Craven. "There's 11 dead now."

Within 24 hours of Dr. Franklin's phone calls that Monday, Dr. David Fraser was in a jet headed for Pennsylvania. He would head a CDC team in what would become the most extensive epidemiologic investigation in the nation's history — the search for the cause of Legionnaires' disease. Before it was over, the outbreak would kill 29 and sicken 153.

***

Ordinarily the CDC is called into an epidemiologic investigation by state health officials, but it took Pennsylvania authorities a while to realize that something was wrong because the first calls from physicians came in over the weekend when most state offices were closed. Five doctors and nurses from four different communities in Pennsylvania had sought assistance on Friday, Saturday, and Sunday.

Dr. Ernest Campbell of Bloomsburg called the regional office of the state health department at about 4 o'clock on Friday afternoon, July 30th. In his hospital were three patients who he thought had contracted typhoid at an American Legion convention. The public nurse who answered the phone told Dr. Campbell that nothing could be done, it was too close to the weekend, when all health offices would be closed.

At almost the very same time, Mrs. Geneva Baxter, the infectious-control nurse for Chambersburg Hospital, was calling the health department office in Chambersburg, 100 miles to the southwest. Mrs. Baxter had three patients who had just returned from the Legion convention and they all had the same symptoms. One of them had died. In this case, the public health nurse tried to help, but by the time the word had passed up and down the ladder — impeded by bureaucratic restrictions on long-distance calls — essential offices had closed for the weekend.

On Saturday, Dr. Howard Hoffmann, the pathologist at Chambersburg Hospital, and Dr. Joseph Brazel of Carlisle Hospital separately called the Philadelphia City Health Department to talk about undiagnosed cases they had, but no definitive action was considered until Monday.

And finally, on Sunday night, Dr. William Ford, chairman of the infection-control committee at Williamsport Hospital, reached a state health official at home. He wanted to tell him about six patients in Williamsport Hospital who had just returned from an American Legion convention with similar symptoms.

It wasn't until early Monday morning that state officials started putting together all the clues that had been coming to them.

A phone call to Edward Hoak, the state adjutant for the American Legion, confirmed their fears.

He said something strange had happened at the Legion's recent convention in Philadelphia because Legionnaires were getting sick and dying all over the state. Hoak said he had been getting calls about sick Legionnaires all weekend, and in fact was just about to call state officials himself.

The information Hoak supplied made it clear that a public health emergency of major proportions was at hand. Immediately the state's investigation was personally taken over by Dr. Leonard Bachman, the energetic, 51-year-old secretary of health.

Dr. Bachman had arrived at his offices that Monday morning with the intention of leaving after a lunchtime executive conference for a vacation at his cottage in Maine. He even had the brass fittings to his sailboat in his jacket pocket.

By mid-morning he had heard that there might be an epidemic in progress. By the time the executive meeting began, the rumor had been confirmed and all efforts were devoted to gearing up for the emergency. By early afternoon, Mrs. Bachman was unpacking their bags.

State laboratories, which would have to analyze samples from patients, were put on around-the-clock status. All public health personnel were alerted to prepare for an extensive epidemiologic survey. The State Police were called in to rush samples to the state labs in Philadelphia by highway patrol and helicopter.

In Philadelphia, which has its own, autonomous health department, the investigation was being put into high gear by City Representative Albert V. Gaudiosi, who, as the mayor's principle advisor, gave the problem the city's highest priority. Gaudiosi had no medical background, but as a former reporter he had a sense for gathering information quickly and efficiently.

Gaudiosi immediately set up a telephone hotline for receiving tips, calls from physicians, and information from people who thought they might have been stricken with the disease. Within three hours after the hotline was set up Monday night, 150 persons had called.

Gaudiosi also called in the police commissioner and had 50 homicide detectives assigned to the case — not because Gaudiosi suspected murder, but because the detectives were trained interviewers who could give the medical people a hand in gathering information.

Between the time Dr. Fraser left his home in the Atlanta suburbs and the time he reached Harrisburg, the number of cases had climbed to 74 and the death toll had reached 19. The figures made him shudder. That was a 25% mortality rate. Few things are that deadly.

The fear among many of the investigators was that this might be the start of the swine flu epidemic that the CDC had been expecting for weeks. The irony that made the fear appalling was that the public health system had anticipated a swine flu epidemic and had even started a $135 million vaccine program, but it would still be months before the new vaccine would be ready.

Dr. Fraser was not able to immediately see the state officials when he arrived because they were in the middle of a press conference, one of the many that were being held to handle questions from the growing numbers of reporters who were converging on Harrisburg. This was just as well because it gave him a chance to talk with the two young EIS officers who had been sent before him — Drs. Ted Tsai and James Beecham.

Dr. Tsai had been rushed to Harrisburg the day before specifically for this investigation. Dr. Beecham, a boyish-looking physician with straight, white-blond hair, was permanently assigned to Harrisburg by the CDC, and this unbelievable week was his first one on the new job.

The two EIS officers had been preparing a "line list", which Dr. Fraser was anxious to see. This was a chart containing 69 bits of information about each patient, including name, age, address, hotel rooms visited during the convention, food eaten, and pertinent medical data.

These lists were mounted on huge boards measuring several feet across so that for each patient a single line could accommodate all the information, enabling epidemiologists to detect common factors by simply comparing one column with another.

Just as a police detective attempts to construct the modus operandi of the criminal he is looking for, so does the epidemiologist try to determine the way an unknown bug or toxin operates. Is it very lethal? Does it spread from person to person? Quickly? How? What is the incubation period — how long does it take to strike its victims down?

For several minutes, Dr. Fraser studied the line lists. The information was incomplete and still preliminary, but one very interesting fact was emerging. None of the cases were new. A lot of the patients had fallen ill or died days ago.

If this terrible disease was contagious, Dr. Fraser thought, they should be seeing a secondary wave of infection by now, especially if it was something as contagious as flu. But maybe the Legionnaires' wives and children and friends were sick at home and hadn't sought hospital care yet. The investigators would have to check that out.

In a nearby room, at the press conference, reporters were pounding Health Secretary Bachman with questions about swine flu. Dr. Bachman would not rule it out, because the results of the lab tests for flu wouldn't be in for two more days.

Fears that this was a swine flu epidemic would persist in the media for the rest of the week, though even now — only 36 hours into the investigation — the people working with the line lists were having their doubts.

Finally the press conference ended, and reporters started pouring out into the health department offices to look at the line lists and to watch the epidemiologists work. Along with them came a young man with a curly, reddish-brown mustache and a ready laugh. He was Dr. William E. Parkin, the state's epidemiology chief.

As soon as Dr. Parkin emerged from the meeting, Dr. Fraser sat down with him and other state and federal officers to map out a strategy. It was obvious that they would have to put together a questionnaire for all investigators to use so that there would be consistent information that could be compared from one patient to another, something more extensive than the line lists already in use. Essentially, what they were looking for was something common to all the victims: A food they had all eaten, a room they had all visited, or even a hallway they might all have hung out in.

It was also very important to determine exactly when the victims were exposed to the pathogen and then how long it took them to get sick. Was it a single exposure on one day, or did it occur repeatedly over the four-day convention? If the Legionnaires had become ill only hours after being exposed to the pathogen, this would suggest a toxin. If it took days or weeks, this would suggest a virus or some other living "bug" that needed time to multiply in the body, until it was strong enough to cause symptoms.

They also had to find out the total number of people exposed — a difficult thing to do, because the American Legion had poor records on the number of nonvoting Legionnaires, family members, and friends who had come to the meeting in Philadelphia. Knowing only the number of people sick or dead was not enough; this would not allow investigators to determine such important things as the disease's attack rate (the percentage of people made sick) or its mortality rate, or whether there was even technically an epidemic.

And eventually, the scientists would have to agree on a definition of Legionnaires' disease, so that patients could be included or excluded from the list of victims on the basis of consistent critera. It would only confuse matters to say, for instance, that everyone who developed a bad cold at the convention had Legionnaires' disease.

Did a person have to have a positive chest X-ray as well as a high fever to qualify as having Legionnaires' disease, or would a high fever and a cough qualify? Did a person have the disease if he had all of the required symptoms but had never walked into the Bellevue Stratford, or would his having come within a block of the hotel be adequate?

The trick was to draw up a definition sufficiently broad to include everyone who had Legionnaires' disease, but sufficiently restrictive to exclude people with random illnesses that had nothing to do with the Legion ailment.

Dr. Fraser realized the importance of planning, but he was also determined to move quickly. He felt that he was in a race with the state laboratories in Philadelphia and the federal labs in Atlanta, which hours earlier had inoculated tissue cultures and eggs with specimens from the sick and dead. Results from these studies would come by Thursday.

The epidemiologists worked all afternoon and into the evening Tuesday on their line lists, the questionnaire and a map of Pennsylvania, into which they were sticking a yellow pin for every sick person and a red pin for every dead one.

By the time that second, hectic day of the investigation was over, there were 106 yellow pins and 20 red ones. To the casual observer, it looked like the epidemic was spreading at an alarming rate, but actually the epidemic had already ended.

Because of late reporting, the epidemiologists were days behind in their statistics. All of the 149 conventioneers destined to get sick were already sick or dead. Of the 26 doomed to die, only eight were still alive.

***

The laboratories in Philadelphia and Atlanta had been working continuously since Monday and would continue to do so throughout the week, driven by the fear that they were looking for a contagious agent that would continue to cause sickness and death until it was identified.

The first specimens to reach the state labs in Philadelphia arrived by helicopter from Chambersburg at 5 p.m. on Monday. They were throat washings from a Thomas Payne.

Microbiologist Vicki Taylor, charged with labeling each sample as it came in, wrote on a sticker "E-76-1", which meant that this was the first sample in the 1976 epidemic. Before the list was completed, they'd write "E-76-267."

Each specimen was then divided. One half was used in the state labs and the other half went by special courier to the CDC in Atlanta, where additional testing was being done.

It was an exciting week for the state lab people, normally resigned to the dreary routine of checking blood for venereal disease, or the minced-up brains of bats for rabies. Here was a mystery agent, potentially so deadly that they must take dramatic precautions to prevent themselves or their environment from being contaminated, and all the world was watching them.

Only two people at a time were allowed in the windowless examining room, which was off-limits to all unnecessary personnel. Gowns, masks, and gloves were used and discarded inside the room before leaving. And mouth pipeting was not permitted: Instead of sucking sample fluids up in a glass straw, the technicians must use a hand-held device to transfer small amounts of fluid from one container to another.

Most people thought that they were looking for a virus, probably the swine flu virus.

Since flu viruses are known to thrive in living, fertilized chicken eggs, both the state and federal labs put specimens from the sick Legionnaires into thousands of eggs hastily ordered from suppliers. If there were any flu viruses present, they would multiply rapidly and become apparent within three days.

Specimens were also injected into growth mediums made up of cells from the kidneys of two species of monkeys, human throat cancer, and an aborted human embryo. Most known viruses would grow in one of these mediums, taking anywhere from a few days to weeks to become noticeable.

And finally some material from the lungs of deceased Legionnaires was injected into the bellies and brains of live suckling mice.

Lung tissue was used because the scientists reasoned that if the virus was to be found anywhere in the bodies of the Legionnaires, it would be in the lungs, since the disease appeared to be a respiratory problem. And the best animals were suckling mice because the newborn are particularly vulnerable to bugs.

Though eggs are usually incubated for 72 hours in checking for influenza, the state didn't want to wait the full period. Instead it was decided to crack some of the eggs at 48 hours. Even though this was early, there should be some signs of a virus if any was growing.

The process began at 2 a.m. Thursday. It was a big moment for Dr. Philip Nash, head of the diagnostic virology lab. He had been in the laboratory night and day since the disease was discovered, grabbing a nap now and then at his desk or on the floor of his office. Now the results of all that work were coming in and he knew that if the tests were positive it would be news of worldwide importance.

Lighting up his pipe, Dr. Nash and his technicians started going through the samples that had been treated and were now ready to read. It was not a difficult evaluation and could be done with the naked eye.

If the samples from the 20 disease victims turned the test fluid a diffuse, rosy color, it meant that the Legionnaires had been infected by a virus. If, however, the fluid remained essentially clear, with clumps of red blood cells at the bottom of the test dishes, it meant that probably no viruses were present.

Holding the lab dishes over his head and looking through them to the light, Dr. Nash saw one clump of red cells after another.

His boss, Dr. Jay Satz, director of the division of virology and immunology, double checked the dishes and saw the same thing. And Dr. Vernon Pidcoe, Dr. Satz's boss, also found only clumps. The results were negative. There was no virus.

The next day in Harrisburg, at a crowded press conference with live radio coverage, the announcement of the negative results was made by the governor of Pennsylvania.

***

While the epidemiologists in Harrisburg were hurriedly combing the state in search of secondary infections, the public health people in Philadelphia were looking at the problem from exactly the opposite perspective.

Harrisburg was looking at the Legionnaires as the source of a spreading disease, Philadelphia was considering them as victims of one that had already spread. The Philadelphia investigation was being headed up by Dr. Walter Ornstein, the chief EIS officer in Philadelphia, and Dr. Robert G. Sharrar, the city's chief epidemiologist and a former EIS officer himself.

Their concern was that at that very moment the Legionnaires' disease might be sweeping through Philadelphia without anyone knowing it. The victims might be thinking they had pneumonia or flu; their doctors might be treating them only for such ailments; and meanwhile the deadly mystery disease would be spreading inexorably.

Pneumonia and influenza are not reportable diseases in Pennsylvania, so no central office had statistics to help them spot such an epidemic. The investigators would have to devise a means for doing so themselves. And they would have to do it fast — but they were swamped by the thousands of phone calls pouring in over the city hotline.

The city detectives brought in by Gaudiosi were very effective as an initial phone screen, gathering basic information and separating the medical from the nonmedical calls. But this still left many hundreds of calls that required a physician to evaluate. The callers had to be contacted again and interviewed in detail, and some had to be visited in person.

One problem was that at this point the epidemiologists in Philadelphia were using a very loose definition of Legionnaires' disease, and they were collecting the names of everyone who had a fever and a cough. Most of these cases were probably only routine illnesses, but the investigators didn't want to discard names that they might need later.

Because there was so much material to process, it was more than a week before they were free to check the possibility that the Legionnaires were only part of a much larger epidemic. The question was how to do this. In the first place, no one had any idea whether the cases reported on the hotline represented an unusual amount of illness for the area at that time of the year. There were no figures from prior years to compare with.

Secondly, there was the problem of defining Legionnaires' disease. The symptoms being reported via the hotline — high fever, cough, and pneumonia — were so common that they could mean anything from Legionnaires' disease to a common cold.

But there was a way to get comparative figures. If Philadelphia was experiencing an epidemic, the investigators reasoned, then hospitals in the area should be treating more cases of pneumonia than usual. Hospitals kept figures like this on file.

Six physicians went to Thomas Jefferson, Pennsylvania, and Graduate hospitals and pored through the scribbled medical records of everyone treated in the emergency rooms from July 1 on.

It took the doctors one and a half days to go through all the records, but finally it was done and an epidemiologic curve was constructed by plotting the daily number of cases since July 1. If there was an epidemic, they should see a jump in the curve on a specific day or week, probably around the time of the Legion convention.

The amounts of respiratory disease at the three hospitals did vary slightly from day to day, but the charted curve was essentially flat. There did not seem to have been more illness than could be expected at the three hospitals.

Still, this could be a fluke. It didn't seem likely, but perhaps the victims of the theoretical citywide epidemic had gone to other hospitals.

It would take too much time to go through the medical records of all 60 hospitals in the metropolitan area, but fortunately there were other things that could be done.

First the city investigators checked the death certificates of all the people who had died in Philadelphia just after the convention. And compared the number of flu and pneumonia deaths with those listed in 1975 and 1974 during comparable weeks. There was no substantial difference.

This wasn't conclusive, because death certificate diagnoses tend to be vague, but it was something.

Then they checked the hotels. The rationale here was that maybe, for some reason, Legionnaires' disease attacked only tourists weakened by the stress of travel and strange surroundings and food or some other unknown thing.

The physicians compared the rate of illness among the guests staying at the Bellevue Stratford with that at the Benjamin Franklin (headquarters of the Legion auxiliary), the Warwick, and the Holiday Inn on City Line Avenue, which is on the outskirts of the city, to see if a suburban location made a difference.

Again the results were negative. The only significant cluster of disease was found at the Bellevue Stratford during the week of July 18, when 15 of 180 persons surveyed reported symptoms similar to those of Legionnaires' disease. This was expected, but none of the other hotels had any statistically significant disease during any of the four weeks surveyed.

Another interesting discovery was that even the illness found at the Bellevue Stratford had occurred only during the week of July 18. Nothing was reported for the week preceding or the two weeks following the week of July 18, indicating that whatever it was that caused Legionnaires' disease came and went between July 18 and 25.

Everywhere the epidemiologists looked for evidence of a citywide epidemic, they found nothing. The only exception was the list of emergency room diagnoses routinely furnished to the health department by city hospitals.

Picking the 11 closest hospitals to the Bellevue, the epidemiologists discovered that 35 cases of pneumonia and flu had been diagnosed during August. This was almost twice the number for the same month in 1975 and 1974.

That would have been a disturbing discovery except for one thing. The "epidemic" suggested by these figures started on Monday, August 2, the day when news of Legionnaires' disease had broken into the media. Before August 2 the number of cases seemed normal for the 11 hospitals.

If everyone was the victim of the same epidemic, the statistical increase should have come during the week of July 21, when the Legion convention was in town.

What apparently had happened was that people frightened by the publicity had flooded the emergency rooms with minor problems that at other times would have gone by unnoticed.

With that factor accounted for, the information gathered by all the studies was consistent. The amount of upper respiratory disease during this period in Philadelphia was no greater than usual. There was no citywide epidemic.

Strange as it might be, Legionnaires' disease seemed to be primarily limited to people who took part in the American Legion Convention. But why?

***

For 18 days, the epidemiologists and other specialists gathered data in Pennsylvania. They pursued a lot of leads that went flat. Finally, the CDC people decided that they could do no more there. On August 20 they returned to Georgia. It was exactly a month to the day after the Legionnaires had started arriving in Philadelphia for their state convention.

The departure of the federal investigators was not itself an admission of failure, because many of them would continue to work on the epidemic back at the CDC in suburban Atlanta. But they returned home resigned to the possibility that the mystery would never be solved.

Any hope for a solution was contained in the millions of bits of information they were taking back with them for a computer to manipulate, looking now as at first for some common thread among the Legionnaires who got sick.

Like Sherlock Holmes going over the scene of the crime on his hands and knees, the epidemiologists pored over their data. But instead of a magnifying glass, their tool was the so-called "ultimate survey" and their method was computer manipulation.

The "ultimate survey" was 10,000 copies of a seven-page questionnaire sent to every one of the 1,002 Legion posts in Pennsylvania. They got back answers from 3,080 persons, including 70 percent of the 2,274 voting delegates at the July convention.

Week after week Dr. Fraser and his people refined their definitions and the group of people being studied, getting rid of as many confounding factors as possible. Soon the stack of computer print-outs was topping six feet

Some interesting facts emerged from the mounds of data.

They found out, for instance, that 17 of the people who got sick had visited the hotel on only one of two days — July 22 and July 23. By counting the number of days from the single visit to the onset of symptoms, the scientists were able to more precisely determine the incubation period of the disease. The incubation period, it turned out, ran from three to eight days.

They also found out that the later a victim arrived, the later he got sick, suggesting that there was a common source of infection and that it hit people when they first appeared at the convention.

In the beginning, the epidemiologists had been comparing all of the people who got sick with other people in the area who didn't get sick, but there were just too many variables between the two groups.

So they reduced the population under study to compare the 149 conventioneers who got sick with the other conventioneers who didn't get sick. But still there were too many variables. For one thing, voting delegates tended to be older and to have done different things from conventioneers who were just visiting.

So the subset was refined even further; the 125 voting delegates who got sick were compared to the voting delegates who did not get sick.

And the final refinement of all was to compare only the 62 voting delegates who stayed in the Bellevue Stratford and got sick with the voting delegates who stayed at the Bellevue Stratford and didn't get sick.

Having so carefully defined the study population for the final computer comparisons, now it was time to get all the pertinent information possible on each person involved and feed that into the computer.

Fourteen EIS officers, all physicians, worked throughout a late September weekend, meticulously going through the charts of Legionnaires' disease patients and extracting from the scribbled pages 800 data bits on each patient for comparison.

It was an exhaustive review, including everything from the number of years the victims smoked, if they did, and whether or not they took Librium, Valium, or other drugs, to the types of treatments that their physicians had used in trying to combat the disease.

The scientists listed whether or not a patient had been given artificial respiration, and if so, for how long. They listed the results of blood, sputum, urine, and stool cultures. They detailed body chemistry in all its varieties for all the patients.

This information was combined with data obtained from the ultimate survey; answers to such strange questions as:

''Can you estimate in hours or fractions of hours how much time you spent on the sidewalk in front of the Bellevue Stratford each day?''

''Did you use the restroom in the lobby during the convention?''

''Were you close enough to touch any pigeons in the area of the Bellevue Stratford Hotel?''

''Can you estimate total time spent in hospitality rooms on each day of the convention?''

Having thus programmed the computer, they then started asking it questions, forcing it to make a myriad of comparisons.

Slowly the picture of a true Legionnaires' disease victim came into focus.

The mean age of the victims was 55. The mean age for those who died was 64, indicating that age was a risk factor.

There was a surprisingly clear and direct association of illness with age; an older Legionnaire was more likely to get sick than a younger one.

People over 70 were four times more likely to get sick than those under 40. This increased risk might be explained by the fact that older people tend to be more vulnerable to disease than younger people. But in this epidemic, there was another consistent variable.

Older Legionnaires tended to be delegates, and the delegates' attack rate of 6.8% was almost 17 times that of nondelegates. What was the explanation — a person's age or his delegate status?

Delegates were more likely to have attended Legion functions; they were the only ones who voted at the meetings; they were more likely to have gone to the hospitality suites, and they were more likely to have hung around in the Bellevue's lobby and hence to have drunk hotel water.

One of the most intriguing associations of all was with the Bellevue Stratford lobby, which was a major area for socializing among the Legionnaires, especially for the delegates who liked to keep in touch with their constituents.

According to the ultimate survey, the sick delegates had spent almost twice as much time in the lobby as the delegates who remained well — 259 minutes as compared to 142.

Maybe it was something in the lobby, but why didn't any of the 36 hotel employees who work only in the lobby get sick?

The scientists had nothing they could put together in a nice, neat package and say: "Here, these are the things that might have caused Legionnaires' disease."

***

While the epidemiologists in Atlanta were poring over their case control studies and the virologists were vainly attempting to grow out the elusive microbe, a pathologist working in a lab only a few blocks north of the Bellevue Stratford Hotel was actually looking at the Legionnaires' disease bug through her electron microscope.

The pathologist was Dr. Sheila Moriber Katz, now professor of pathology and laboratory medicine at Hahnemann University. Working with lung biopsy material taken from a 50-year-old Legionnaire stricken with the mysterious disease, Dr. Katz had visualized unusual oval-shaped structures with outer membranes.

Writing later in the *Archives of Pathology and Laboratory Medicine,* she and Dr. Isador Brodsky, director of Hahnemann's cancer institute, would say they thought they might have detected mycoplasma, chlamydia, rickettsia, small intracellular bacteria, or an unrecognized microorganism.

The two Hahnemann physicians immediately brought their findings to the attention of other pathologists and the CDC investigators. But discovering the unusual structures was a long way from proving that they caused Legionnaires' disease, so the CDC never followed up this lead.

A few months after starting work with the diseased lung specimens, Dr. Katz herself came down with pneumonia. Immediately suspecting that she had caught Legionnaires' disease, the 33-year-old pathologist took blood and sputum samples from herself for later study. Subsequent blood tests would prove negative, but sputum analysis would suggest that she might have been infected with the Legionnaire microbe.

The CDC scientists kept working on the mystery until the end of the year, but the investigation was winding down. Epidemiologists and lab personnel were switched to different assignments and only a handful of people continued to work on Legionnaires' disease. But even in December, at the lowest point of all in the investigation, when the CDC was at the point of conceding failure, the young Dr. Fraser continued to be energetic and enthusiastic about the work.

Dressed in his workman's boots and open-necked business shirt, he went from his office, checking with the dwindling number of colleagues still working on the project. At the height of the investigation, a couple of hundred experts had been studying the mystery. Now there were only a dozen or so, and these were working part-time, more out of personal curiosity than as part of a work assignment.

Dr. Fraser will never know why he was so insistent on seeing to it that everyone received a copy of the EPI-2 he had just completed. This was the second epidemiologic report on the declining investigation.

Probably more out of a passion for completeness than anything else, Dr. Fraser took it around to all the labs that had been working on the mystery.

Reports like these are standard at the CDC, keeping everyone informed about what everyone else is doing in a given study.

An epidemiologic investigation as big as the one on Legionnaires' disease involves hundreds of people in the laboratory and in the field at one point or another. Many of them work essentially independently, studying tiny pieces of the puzzle, and never come into contact with their colleagues working elsewhere. They never know when a scientist next door, or down the hall, or up in Philadelphia might have uncovered information that is essential for them to progress in their work.

Grabbing several of the freshly printed reports that his secretary had just brought in, Dr. Fraser bounded out of his fifth-floor office in Building One, took the elevator to the first floor and walked across the open-air walkway to Building Seven, which houses the infectious-disease laboratories.

While he was there, he decided to drop off a report at the leprosy rickettsia laboratory of Dr. Charles C. Shepard. It had seemed unlikely that Legionnaires' disease was related to leprosy or caused by rickettsia, a microorganism that causes such things as spotted fever and is usually insect-borne. But in August, Dr. Shepard's laboratory had taken a look at the blood samples just to make sure.

Dr. Fraser's compulsive thoroughness in seeing that Dr. Shepard's lab received a copy of the EPI-2 would pay off handsomely. One of the people who would browse through the report would be microbiologist Dr. Joseph McDade.

***

During the first few weeks after the outbreak in Philadelphia, Dr. McDade had been responsible for checking specimens for the presence of rickettsia. It was a long-shot and indeed Dr. McDade found no evidence of this microorganism at the time.

So early during the investigation he filed the microscopic slides away in a wooden box on his desk and moved on to other work at the CDC. He didn't become involved with Legionnaires' disease until that day in December when he came across Dr. Fraser's EPI-2 report.

Idly thumbing through the report, Dr. McDade came across a reference to the fact that the SGOT levels in some Legionnaires' disease patients were elevated. Elevation of the enzyme SGOT indicates liver damage and is frequently, though not exclusively, associated with rickettsial infections.

Suddenly, the words and figures seemed to jump out of the pages at him. "What if this was a new group of rickettsia?" Dr. McDade asked himself as he stared at the report. It was a presumptuous speculation, and Dr. McDade knew it, because no new type of rickettsia had been identified in more than 40 years.

Still, there was always the possibility, Dr. McDade thought as he reached across the desk to the wooden slidebox. What harm would it do to examine the slides again now that he had the leisure to do it? This time, instead of looking only for rickettsiae of the usual size and shape, he would pay particular attention to anything unfamiliar, in case it might be a new rickettsial form.

This probably wouldn't lead to anything, Dr. McDade thought as he began the tedious process of reviewing the slides. But it was a slow day, and he didn't have anything better to do.

***

Dr. McDade saw nothing in the first few slides he reviewed. Staring at each one for 20 to 30 minutes, he slowly manipulated the slides under the microscope, carefully examining the myriad patterns, looking for something that resembled a rickettsia or a bacterium.

By 11 a.m., he had found what he was looking for. The bugs almost popped out at him as he brought them into the center of the microscopic field.

There was a group of bright red rods within the purplish blob of a spleen cell. They seemed to be the same kind of long, fat rods he had seen in August — but those had been few and scattered, while these were clustered, which indicated a serious infection.

Dr. McDade spent the next hour going over more slides, but rushing excitedly through them now. Then he called in his boss, Dr. Shepard, who knows as much as or more about rickettsia than any other scientist in the world.

Dr. McDade wanted an unbiased evaluation of what he had found. So, without telling Dr. Shepard where the slide had come from, he merely stood aside as his boss stared into the microscope.

''That's the most impressive thing you've hit yet,'' Dr. Shapard said, thinking that it was a slide from a spotted-fever case that Dr. McDade had been working on. ''It looks like it might be something.''

Still, Dr. McDade mentioned nothing about Legionnaires' disease, and Dr. Shepard left, unaware of the historic discovery that was taking place in his laboratory.

***

Sitting alone in his crowded office, the data from the hectic work he had done months earlier scattered about him, Dr. McDade decided to try again to grow the Legionnaire microorganism. Months earlier he had attempted to do this, injecting specimens from Legionnaires' disease victims into guinea pigs whose spleens he then studied. He assumed the lungs would contain the bug, if there was one, because all human victims of the disease had pneumonia. But the results were negative. That was because the spleens of guinea pigs are very effective in screening out bacteria before they reach the lungs.

This time suspecting the possibility of bacteria or an unusual form of rickettsia, Dr. McDade decided to study tissue from the spleens of guinea pigs inoculated with specimens from deceased Legionnaires' disease victims. Also in the earlier studies, antibiotics were used to protect the samples from bacterial contaminants, but this time Dr. McDade took care to keep antibiotics from the specimens. If this strange bug he had seen was a bacterium or an unusual form of rickettsia vulnerable to antibiotics, he didn't want to inadvertantly kill it with antibiotics. Dr. McDade even went so far as to order special eggs from flocks of chickens that hadn't been given antibiotics with their feed, a common practice in egg production. It took three days for the eggs to come in, but they finally arrived. And, on December 30, Dr. McDade injected them. If he was on to anything, the chicken embryos should start dying within a week.

The next few days were anxious ones for Dr. McDade. He started each morning by examining the eggs over a light to see if the embryos inside were alive or dead. Nothing happened for the first few days — but then, on January 5, six days after inoculation, the embryos started to die. By January 6, most of them were dead.

Dr. McDade was sure he had the bug. Quickly he made slides from the egg material, stained them and put the material under the microscope. The microscopic field was littered with clusters of red rods. The eggs were filled with the strange organism.

Trying to contain his excitement and appear the cool scientist, Dr. McDade told Dr. Shepard and a couple of colleagues in the small laboratory. What a secret they shared! But they couldn't tell anyone outside the laboratory until they were more certain. It was clear that Dr. McDade had found something. But was it the same microorganism that had caused Legionnaires' disease? There was one way to tell. Did the victims' blood contain antibodies to this bug?

***

Dr. McDade would determine this by mixing the bugs he had just grown with blood from Legionnaires' disease victims. If the victims had been infected by this bacterium, the antibodies in their blood would attack it, a reaction that the laboratory could easily detect with an indirect fluorescent antibody (IFA) test.

In an IFA test, the slides are treated with a substance that fluoresces under ultraviolet light if the antibodies present react with the bacteria. Using his influence, Dr. Shepard got four samples of the precious Legionnaire blood from the virology lab on the floor above without explaining why he wanted it.

It took a while to prepare the slides, so it wasn't until almost 4 o'clock on January 7 that the scientists were ready to look at them under the ultraviolet light. Snapping off the overhead light in the small microscope room, they put a slide under the microscope and turned on the ultraviolet light.

Blotches of eerie green fluoresced under the microscope. The test was positive. There was an antibody reaction. This particular victim had been infected with the bacteria that Dr. McDade had isolated.

Quickly the scientists tested the sample from the other three patients. From one, there was fluorescence. But the two other samples were negative. Could it have been a fluke? Dr. Shepard rushed upstairs to get more samples before the CDC closed for the day.

These were paired samples, meaning that two blood samples had been taken from each patient. The first was drawn during the acute phase of the illness, soon after the patient was infected and before his body was able to develop immunity by producing antibodies. The second blood sample was obtained weeks later, during the convalescent period, when the blood contained millions of the antibodies it had made to fight the bug. If the second sample in the paired serums showed a sharp rise in antibodies over the first sample, it would constitute convincing evidence that the scientists had the right bug.

Once again the samples were prepared by Martha Redus, a microbiologist, and Dr. Verne Newhouse, an insect specialist, who were more experienced with the IFA test than anyone else in the laboratory.

It was just as well that they were running the tests, because Dr. McDade was too exhausted and excited at this point to do anything constructive. So he went home. Miss Redus would call him as soon as they had something.

Dr. McDade had little appetite when he got home that night and couldn't concentrate on anything. All week long he had been having trouble sleeping. His wife didn't know what was wrong because he hadn't told her about what was going on in the lab.

Finally, at 8 p.m. on June 7, the telephone rang in the McDade home. It was Martha Redus. Two of the five paired serums showed a sharp increase in the antibody level, she said.

The test was holding up. It was becoming more and more likely that Dr. McDade had found the bug that had caused Legionnaires' disease.

What was needed now was numbers. Was this test specific for Legionnaires' disease, or would it come up positive with any pneumonia case, even those caused by viruses, other bacteria or rickettsia? If it was nonspecific, if it always came up positive, then it would be worthless.

The scientists tested serums from patients known to have had viral pneumonia. The results were negative. They tested serums from persons caused by rickettsiae. These tests were negative, too. Serums from patients with psittacosis, a pneumonia caused by a fungus, were tested. Negative.

But time after time the samples from Legionnaires' disease patients came up positive.

By the middle of January, Dr. Shepard was sure enough about what his lab had found to call Dr. David J. Sencer, the head of the CDC, and request a conference.

The meeting was held on January 13 in Dr. Sencer's spacious, second-floor office. Among those present was Dr. Fraser, head of the epidemiological investigation of Legionnaires' disease. It was the first time that Dr. McDade had met him.

Drs. McDade and Shepard quietly outlined the data they had accumulated over the previous week, without drawing any conclusions. They didn't have to. The data were spectacular. Their significance was obvious.

The scientists might have been expected to call a press conference and proudly announce the breakthrough. But everyone in the room that day was still too wary. Already too many scientists around the country had announced promising theories, only to retract them a few weeks later.

What with the swine flu fiasco and the frustrations of Legionnaires' disease, the CDC was having too much trouble with its credibility to want to jeopardize it further.

But conversely, with an epidemic as well publicized and frightening as Legionnaires' disease, the CDC couldn't hold back important information long. As more and more people in the CDC discovered what was happening, the press would be more and more likely to hear about it.

The laboratory moved quickly now to amass as much information as possible, knowing that it was only a matter of days before some kind of announcement would be made. If the new bacterium was going to prove a ringer, let it be now before the announcement, everyone prayed. Dozens more serums were rushed through the IFA test. The results continued to be consistent.

For months Dr. Shepard had been intrigued by the similarity of Legionnaires' disease to whatever caused the 1965 epidemic at St. Elizabeth's, a District of Columbia mental hospital. The epidemic struck 94 persons and killed 16. Now he had a reliable way to determine whether there was a connection. The CDC's policy of preserving blood samples and tissue from disease outbreaks would pay off.

Specimens from the St. Elizabeth outbreak were obtained from the CDC's refrigerated serum bank. No one was told why because Shepard's people were still trying to keep the investigation secret.

Still cold from their long, frigid hibernation, the little vials of serum were prepared by Miss Redus and Dr. Newhouse, tagged with the fluorescent substance, and put under the microscope with the ultraviolet light.

There was no mistaking it. Thirteen of the 14 slides fluoresced with the glow of the bugs that caused Legionnaires' disease.

It was a strange feeling for the people in the darkened room, looking at the greenish glow and realizing that it was produced by the same microorganisms that had caused a lethal epidemic so long ago. For 11 years these bacteria had hibernated in the CDC freezers, eluding detection. But now they had been unmasked.

Another request was sent to the serum bank. This time the scientists wanted samples from an epidemic in Pontiac, Michigan. It seemed less likely that this epidemic would be linked to the Legion bug, because the disease had seemed so different. None of the 144 persons who were stricken died and there was no pneumonia.

But the scientists ran the tests anyway. About 90% of the serums were positive.

It was amazing. In one sweeping move the CDC had cracked three of its most vexing unsolved cases. And all because Dr. McDade was curious about something mentioned in a technical report on a day when he had nothing better to do.

***

An announcement was scheduled for January 18, much against the wishes of Dr. Shepard and other scientists, who were not accustomed to high-pressure, celebrity science. They felt that more tests should be run to make absolutely certain that what they had found wasn't a fluke.

The evidence was already overwhelming, but not absolutely definite. Respected scientists gain their reputation by avoiding foolish mistakes and by being so consistently conservative in their conclusions that they are rarely wrong. To err in a case like Legionnaires' disease with a premature announcement could cost the CDC scientists some of their carefully built reputations. But officials in the CDC insisted on making the announcement as soon as possible.

And so on that Tuesday afternoon, in an auditorium filled with 300 cheering CDC employees, Dr. Sencer made the historic announcement.

He used a lot of the vague, qualified phrases that scientists like saying, for instance, that they had found a bacterium ''quite definitely associated with the disease.'' But it was clear to everyone what it really amounted to:

The Centers for Disease Control had found the cause of Legionnaires' disease.

Index

INDEX

A

B

C

D

E

F

G

H

I

J

K

L

M

N

O

P

Q

R

S

T

U

V

W

Y

For Product Safety Concerns and Information please contact our EU representative GPSR@taylorandfrancis.com Taylor & Francis Verlag GmbH, Kaufingerstraße 24, 80331 München, Germany

Batch number: 10731074

Printed by Printforce, the Netherlands